DIGGING BY STEAM

The massive Darby broadside digging engine at work in the fens. This 8 H.P. Savage built machine was owned by J. W. Moss
(Major Ind collection)

Digging by Steam

A History of Steam Cultivation by means of the application of Steam Power to the Fork, Mattock and similar implements

by

Colin Tyler
M.I.W.M.
Member of the Newcomen Society

MODEL & ALLIED PUBLICATIONS
Argus Books Ltd,
14 St James Road,
Watford, Herts,
England

© 1977 Argus Books Ltd

Text © 1977 Colin Tyler

All rights reserved. No part of this book may be reproduced in any form or by any means without the permission of Argus Books Ltd.

First published 1977

ISBN 085242 522 8

Printed in Great Britain
by W & J Mackay Limited, Chatham

"I say the plough has sentence of death passed upon it, *because it is essentially imperfect.*"
C. W. HOSKYNS, *Talpa*, 1853.

By the same author and John Haining:
"Ploughing by Steam"

To my wife and daughter, Everett and Sheena.

Contents

Chapter		Page
	FOREWORD	9
	INTRODUCTION	13
	Illustration Index	15
1	ROTATION OR RECIPROCATION?	19
2	AMBULATING BROADSIDERS	33
3	VEHICULAR BROADSIDERS	49
4	LAST OF THE BROADSIDERS	63
5	DARBY TENDER DIGGERS	82
6	TENDER DIGGERS—OTHER TYPES	93
7	HOP DIGGERS	114
8	EARLY ROTARIANS	128
9	LATER ROTARIANS	139
10	ARCHIMEDIAN DIGGERS	159
11	DEEPER DIGGINGS	164
	Broadside digger details	
	Digging engines made	
	Patents list	
	Acknowledgements	
	BIBLIOGRAPHY	169
	INDEX	171

Foreword

BEING ONE OF THE PEOPLE to whom unorthodox and unusual Agricultural engines and machinery have always presented great interest, none more than the products of small country workshops and yards of long ago, the engines described and illustrated in this book and in particular those magnificently complex machines the broadside diggers form a subject of instant and irresistible appeal.

Fresh in my own memory are the long hours spent in researching through old records, letters, notebooks and drawings when preparing material as co-author with Colin Tyler, of the companion volume to this book *Ploughing by Steam*.

As the son of a farmer, brought up with and trained in steam and with a strong agricultural engineering background this probably presented somewhat less of a problem to me than it might have done, but to seek out and collate facts relating to machines built in limited numbers so many years ago, not one of which now survives even in part, as the author of this book has so faithfully done, is surely a task of Herculean proportions, deserving of more than ordinary commendation by all those whose interest is vested in mechanical and agricultural history.

Why did the principle of fork or rotary action not survive, and why the lapse of time before the rotary action re-emerged in the guise of the big Fowler Gyrotillers, with rear mounted vertical tines, in the twenties, followed by the horizontal rotary actions standardised on the farm Rotovator and a whole host of horticultural machines since the Second World War?

Obviously the earlier machines such as Ushers or Ricketts, were underpowered for the work expected of them and suffered severely from imperfections in the materials from which they were constructed and in some cases from poor detail design as well.

One cannot but wonder too, if some of these early pioneers had fully grasped the principles of soil cultivation, later to be clearly defined by Professor W. R. Williams in his *Principles of Agriculture* where he states that, while the main reason for cultivating soil is to produce a "crumb structure" in the arable horizon, if all the interstices between the "crumbs" are filled with dispersed soil particles the soil will behave indistinguishably from a structureless soil, and the agricultural value of a soil structure lies in the fact that there are wide "pores" between the aggregates formed by the cementing together of a mass of structureless soil particles. If these pores become filled with a finely divided unaggregated soil all the biological processes and movement of water will take place as though in a dispersed structureless medium. Consequently the important addition to the definition already quoted, of the purpose of soil cultivation, should be made that cultivation must produce a crumb structure throughout the whole of the arable horizon, but in such a way that the smallest possible quantity of dispersed soil particles is created at the same time; a soil 65% aggregated and 35% dispersed will not in fact differ from a soil 100% dispersed.

From these deductions it will be seen therefore that the first requirement that a cultivator of any type must fulfil is that the creation of a structural condition in the soil should be accompanied by the lowest possible degree of dispersion and although a soil will lose its crumb structure and stability after it has been cultivated and sown with an annual crop through physico-chemical and biological causes, it is essential to reduce to the minimum any mechanical cause of loss of structure.

These theories probably put into words what many arable farmers of the last century, accustomed to fairly shallow cultivation of the soil by horse drawn implements already knew, albeit in different phraseology: some prejudice against the double engine system of cable ploughing remained for many years particularly in areas where over-deep ploughing had inverted the subsoil and ruined hitherto productive fields and possibly the thought lingered still in the minds of many arable farmers that the broadside and other diggers might do the same disservice to the land.

Fowlers of Leeds, doyen of steam cable cultivation, were careful to point out that their heavy balance knifers, for instance, should be used—as they were extensively abroad—to aerate the subsoil and *not* invert it, and if inversion of the sandy subsoil deep below a thick clay pan was required to mix it with the topsoil, the heavy, slow heathplough should be used. This technique was used abroad but not generally in these islands.

Perhaps another reason that the diggers, particularly the broadside machines, never became really popular was the amount of work necessary to set them down to work. Cable ploughing was never itself a light job and sitting on a cultivator for ten, twelve or even longer hours a day never came up to the standards of comfort enjoyed by the driver of a present day tractor with all enclosed weather proof safety cab and all "mod. con.": but a fair amount of heavy work must have been necessary to move "The Enterprise", for example, through narrow roads and lanes—no low loaders in those days!—through gates into fields and so into position for broadside working. It is a rural fact of life even today that few field gates are sited to facilitate entry or exit by large unwieldy machines and this situation must have been even worse seventy or eighty years ago. We can but conjecture at the lack of countryside popularity of these fine steam machines and perhaps it would be fairer to lay the blame, if any existed, on bad publicity and lack of advertising finesse. Leaving aside any doubts as to the effect on the soil and additional loading on the forks of passing a heavy engine over ground to be worked, all the diggers described in these pages were wonderful examples of original thinking, engineering skill, and ingenuity; the later Coopers in particular displayed some excellent design work with that special look of "rightness" so sadly lacking in some machinery of this strident modern age.

What would I not give to be able to lift the dusty curtain of the years, however briefly and see, hear and smell the big broadside digger as it slowly traversed that sunny Essex field, a faint trail of smoke from each chimney and the lightest feather of steam showing from the safety valves as the forks move in their set rhythm across the wide work strip. Though, so far as is known, no diggers now exist, the author's superb one-sixth full size model of the Savage 6 N.H.P. broadside digger bids fair to fill the gap. This engine with all the intricacies of the full size machine reproduced faithfully in full detail and the original works drawings followed to the last detail, will make it possible to realise fully the tremendous amount of thought and ingenuity expended on the design and construction of the original engine.

Not only, therefore, has Colin Tyler brought to life a digger for our delight, but his long standing and abiding interest in diggers of all types and his deep and conscientious researching into the subject has enabled him to record for posterity, with a thoroughness that will confirm him as an authority on a chapter in agricultural history that might well have remained closed and even, eventually lost for ever.

Tredunnock, Usk JOHN HAINING

Introduction

THE GREAT NUMBER of books published in recent years dealing with the steam engine in its various forms may—understandably—make the reader wonder why yet another has appeared, particularly when one considers that apparently every conceivable aspect connected with the history of the steam engine has been adequately covered. I make my excuse for writing this book by making a somewhat unusual claim for it in that it is concerned with a type of agricultural steam engine which has so far apparently escaped the attention of historians and steam enthusiasts alike, an omission that I hope will be corrected by the information set down here. It is as complete as possible in the light of information discovered so far, and gathered from many sources both private and public. I hasten to add that it is one thing to claim writing the first words on a subject, but it is quite another to claim having written the last words.

It is strange that steam digging engines have not received attention before, as among their numbers are included some of the most intriguing and bizarre designs of steam engines ever to be made during the reign of Queen Victoria.

Most of the illustrations have not been published together before, but may have appeared as individual illustrations in various publications over the last one hundred and thirty years. In many cases only one or two pictures or drawings exist of each type of engine and a few are known only by description, sometimes in vague outline.

As with any volume of this nature, it would have been impossible to write it without a great deal of assistance which was always given unstintingly by all of whom information was asked.

Special acknowledgement must be made to Arthur Johnson who spent most of his working life constructing many types of steam engines with Savages of Kings Lynn, retiring as Works Manager of that now sadly defunct Company. It is entirely due to Mr. Johnson's help in providing drawings and details of the 6 H.P. Savage built Darby broadside digger that it was possible for me to construct the two inch scale model featured in the following pages, which at the time of writing is unique in being the only known digger in existence—albeit in miniature.

The rare illustrations were obtained from many sources including Record Offices, Libraries, Museums and private individuals to all of whom I am greatly indebted. My thanks also to John Creasey of the Museum of English Rural Life for his endeavours in providing photographs from the Museum's archives and to Everett, my patient wife, for typing this volume from my illegible manuscript.

I believe that the largest number of digger illustrations assembled to date are contained in this volume, but the variety of types and numbers of diggers shown is not great when compared with, say, ploughing engines. The search for new material is continuing, therefore should any reader have any new information on steam diggers whether in printed or picture form the writer would be most interested to learn of it.

Studies of technical history can be adventures for the individual. A detective hunt into the past, exploring the many avenues of the development of machines, each avenue leading to exciting discoveries which when pieced together combine to make one of the most intriguing stories to emerge from the fascinating Victorian era. It is the sum result of these researches which are related here and it is my earnest hope that it will not only be of interest but an enjoyable view of one aspect of Victorian agricultural engineering.

Tilehurst, Reading C. R. T.

Illustration Index

Fig. No.		Page
	"The Enterprise" broadside digger	Frontispiece
1	Rotary cultivation in the Industrial Revolution	19
2	Fork digging	20
3	Spade digging	21
4	Fowler 14 H.P. ploughing engine	22
5	Comstock's rotary spader	24
6	"Talpa" cartoon	27
7	Rickett's cultivator	28
8	Nubar's cultivating engine	30
9	Darby broadside digger	31
10	Darby broadside digger	33
11	Darby walking digger	36
12	Darby walking digger	37
13	Darby walking digger	38
14	Darby walking mechanism	39
15	Darby walking digger	40
16	Darby walking digger	42
17	Darby walking digger	43
18	Darby's workshops	45
19	Darby's catalogue cover	46
20	Darby broadside digger	49
21	Agricultural and General Engineering Co.'s advertisement	50
22	Agricultural and General Engineering Co.'s digger	51
23	J. and H. McLaren broadside digger	52
24	Broadsider in road setting	53
25	Broadsider at work	59
26	Savage broadsider at work	63
27	"The Enterprise" working	65
28	"The Enterprise" working	65
29	Frederick Savage statue	66
30	Savage 6 H.P. broadsider working	71
31	Savage's machine shop	74
32	2 in. scale Savage broadsider model	74
33	2 in. scale Savage broadsider model	76
34	2 in. scale Savage broadsider model	77
35	2 in. scale Savage broadsider model	77
36	2 in. scale Savage broadsider model	78
37	2 in. scale Savage broadsider model	79
38	Savage pattern warehouse	80
39	Darby "Quick Speed" digger	82
40	Darby "Quick Speed" digger	84

ILLUSTRATION INDEX

Fig. No.		Page
41	Darby "Quick Speed" digger	85
42	Darby "Quick Speed" digger	85
43	Darby "Quick Speed" digger	86
44	Darby "Quick Speed" digger	86
45	Darby and Steevenson digger "Nottingham"	87
46	Darby and Steevenson digger "Colchester"	88
47	Darby and Steevenson digger "Colchester"	89
48	Darby and Steevenson digger "Colchester"	90
49	King George V views digger	91
50	Fowler "Gyrotiller"	91
51	Cooper traction engine	94
52	Cooper traction engine	95
53	Cooper digging mechanism	95
54	Cooper digger	97
55	Fowler-Cooper digger	98
56	Fowler-Cooper digger	98
57	Cooper digger	99
58	Cooper digger at work	99
59	Cooper digger cylinders	100
60	Cooper undertype digger	102
61	Cooper undertype digger	102
62	Cooper No. 5 digger	103
63	Cooper No. 5 digger	103
64	Cooper chain drive digger	104
65	Cooper chain drive digger	104
66	Garrett digger	106
67	Nagy digger	107
68	Nubar digger	107
69	Garrood digger	108
70	Proctor digger	109
71	Proctor digger	110
72	Burrell-Proctor digger	111
73	Pryor's mechanism	112
74	Knight hop digger	114
75	Knight-Howard hop digger	116
76	Knight-Hetherington and Parker hop digger	118
77	Knight-Hetherington and Parker hop digger	118
78	Knight-Howard hop digger	120
79	Knight-Howard hop digger	120
80	Knight hop digger	122
81	Knight hop digger at work	122
82	Knight-Howard hop digger	123
83	Knight-Howard hop digger	123
84	Knight's digger mechanism	124
85	Knight's rope drive	126
86	Knight's system	126
87	Trevethick "Tormentor"	129
88	Trevethick engine	130
89	Pratt digger	132
90	Lillie digger	135
91	Barrat's "la piocheuse"	136
92	Barrat-frere digger	137
93	Keddy digger	138

ILLUSTRATION INDEX

Fig. No.		Page
94	Usher cultivator model	140
95	Usher cultivator	141
96	Rickett cultivator	143
97	Rickett mechanism	144
98	Rickett cultivator	146
99	Rickett cultivator	146
100	Rickett steam carriage	148
101	Bethell cultivator	149
102	Bethell cultivator	150
103	Romaine horse drawn cultivator	152
104	Romaine-Crosskill cultivator	154
105	Romaine-Nash cultivator	155
106	Hoskyn cultivator	157
107	Holcroft rotary engine	159
108	Monckton and Clark cultivator	160
109	Parker cultivator	161
110	Brennand cultivator	162

Fig. 1. A composite engraving of the Industrial Revolution. Agriculture is represented not by a plough, but by a rotary cultivator
(*International Exhibition Catalogue, 1862*)

CHAPTER ONE

Rotation or Reciprocation?

THE SOLUTION OF THE apparently simple task of cultivating soil has occupied the minds of farmers and mechanics alike for generations.

From the beginning of civilisation until the present time the plough has been pulled through the soil by various means from humans to animals and then by machines.

The first ploughs merely scratched the soil and after a long period were redesigned to include a share and mouldboard, turning the soil over leaving in its wake familiar lines of neatly turned furrows which have paraded across the land for centuries.

Even older than ploughing is a humbler, less romantic but very useful process which today is perhaps more familiar to gardeners than farmers—namely, digging using the fork or spade.

Early in the nineteenth century, ploughing and manual digging were commonly practised and it followed that comparisons were made between the quality and cost of resultant crops obtained from the soil. Arguments raged back and forth as to the relative merits of each method.

It was said that although less expensive to plough than dig (Fig. 2) the very action or reaction of mouldboard against underlying soil produced a hard pan, impenetrable by roots. Digging when properly done by labourers produced a shallower tilth, without the pan and more suitable for a seed bed. However the physical advantages of manual digging were outweighed by the financial advantage of ploughing. Digging was much more expensive due to the high labour content involved. Additionally sheer hard manual work was performed

Fig. 2. In the early 19th century, hard labour on the land was a matter of course. It was the human action of digging that the designers of some digging engines attempted to imitate. An uncanny similarity was achieved, particularly with the broadsiders
(*Museum of English Rural Life—cited M.E.R.L. hereafter*)

by labourers who tended to skimp the work and leave the surface apparently well dug, but investigation would reveal the top surface only was disturbed to a shallow depth, not the full spit required.

Ploughing could be seen to be well done, digging could not. It was the difference between a job which could be done well by horse power and badly by manual labour.

Mortimers *Art of Husbandry* illustrates the cost of manual digging in 1708. He based his estimates on a wage of fourteen pence a day in summer and twelve pence a day in winter—about 6/6. (32½p) per week.

He says "Of Digging of Land. A man may dig four or five square pole in a Day of Garden-gound, a Spit deep, that hath been dug already; and he may dig three Pole or more to fill into Wheel barrows and something better than two Poles into Carts, if 'tis good digging Ground.

"The common Price for digging of Garden-ground is Fourpence a Pole, because they expect better Wages for Garden, than other Work, else Threepence a pole is good wages. To dig ground a Spit deep, and to fill it into Wheelbarrows is worth Fourpence a Pole, and into Carts Six-pence a Pole."

Mortimers Supplement of 1712 adds that "To how Hop Grounds is 4s. an Acre, and to make up the Hills is 2s. an Acre.

"To dig Land for a Hop Ground is worth from 3s. to 4s. a Pole square, they reckon that a thousand Hills ought to be planted in an Acre, and six or seven

Plants on each Hill, which plants they commonly sell about us for 8d. or 9d. which they give 10s. an acre to plant".

Observing that there are 160 square poles to the acre, the astounding cost of digging amounted to £2 13/4d. (£2·67) per acre. Further at the rate of five poles per day, it would take one labourer thirty two days to dig an acre.

In 1860 a labourers wage was about 12/- (60p) and in 1895 16/- (80p) a week compared with 6/6d. (32½p) in 1708. It would therefore have cost about £7 to manually dig an acre in the 1870s. Assuming that digging was a desirable form of cultivation, indeed considered essential by some particularly in hop fields, then there was a clear case for developing a machine incorporating the power of steam to replace the expensive, slow and inefficient labourer (Fig. 3).

The steam age was an era of mechanics and machines. No electricity, electronics or petrol. A time of iron and brass, spur and bevel gears, smoke and grime as far removed from us now as the Iron Age was from the Victorians. A period of social inequality, ignorance for the majority and glory for the few with the capability and means to rise above the general poverty, unless lucky enough to be born into the comparatively few "right" families. The new, almost untried power of steam combined the ancient alchemic elements of fire, water and air. In the middle ages it would have ensured an untimely end for any who dared to combine these elements into an unseen force to make machines move without aid. Uncontainable in excess, the very danger associated with steam struck awe into the beholders of the first hissing monsters whether they were pumping water or proceeding jerkily along fish bellied rails.

Fig. 3. The spade as an implement was never used on digging engines. Pressure required to enter the blade into the soil is too great, and obstructions would easily damage it. The fork breaks up the soil as well as tumbling it over, making a better seed bed
(M.E.R.L.)

It so happened that the embryonic steam engine developed unto a useful source of power during the reign of Queen Victoria, the same period during which the British Empire was being built. Size in both mechanical and geographical terms was almost unlimited. From one of Earths smaller islands emerged a culture which brooked no opposition, spreading over the globe's surface to its most distant parts. But back in that little island, men were improving the Eighteenth Century phenomenon of steam power that Savory, Newcomen and a few others had attempted to harness, ill equipped as they were with little knowledge, few tools and poor materials. What giants of pioneering they were, faced with odds against success which would daunt all but the most determined.

Early improvements made on the low pressure atmospheric engines by Watt were not sufficiently revolutionary, achieving increased efficiency on a type of engine that in itself was inherently inefficient in fuel consumption, and too unwieldy for application to locomotion.

More radical thinking was needed and it was not until Murdock, Trevethick and others started exploring the possibilities of high pressure steam that any real chance of comparatively high efficiency from a small engine was realised.

Knowledge was accumulating, each inventor learning from others mistakes. The increase in boiler pressures brought about developments of new boiler materials, eventually replacing the original wrought iron with steels of far

Fig. 4. This magnificent 14 H.P. Fowler ploughing engine of 1870 was one of a pair numbered 1329 and 1332. It was engines of this type that digging enthusiasts were hoping to oust from the fields
(*Major Ind collection*)

greater strength. More accurately machined parts turned out in greater quantities resulted in cheaper manufacturing costs and gains in mechanical efficiency. Thus the names of Bessemer, Maudsley, Nasmyth and other mechanical engineers will always be indirectly associated with the development of steam power.

With the coming of the mechanical age, a school of thought emerged that in general principle attempted to mechanise agriculture by the application of steam power to the land. Here was obvious commercial potential in taking the drudgery and labour out of cultivation. Most inventors turned their thoughts to the plough and in various ways attempted to replace the horse (Fig. 4), a story which is detailed in the companion volume *Ploughing by Steam*.

A few ingenious minds turned in other directions and some considered the application of steam to the as yet entirely manual job of digging—a much more difficult task for the engineer, as he was starting literally with only a humble fork, whereas the plough at least was of established design. A fork could not be pulled through the ground with a cable, it could not be made easy to dig simply by the addition of steam power.

An entirely new concept was required. A machine that would work independently of cables, be a self contained unit capable of performing its work cheaply and efficiently, eliminating the human failure of skimping.

Such were the requirements of a digging machine and the ones who sought a solution were to be hard put in finding it. However, a challenge rarely daunts an inventor and from the early part of the nineteenth century some very strange machines began to appear (Fig. 5).

The means of mechanising an essentially human activity was approached in diverse ways, some of which are discussed in these pages, but it is true to say that digging exponents eventually fell into two broad categories, both attempting to achieve the same object.

Early digging engines employed the rotary principle and were generally a traction type engine equipped either at the front or rear with a horizontally mounted set of rotating spades, forks or knives. These engines form the earliest group and can be said to have worked on similar principles to the present day light internal combustion rotary cultivators.

In the early 1800s though, it was not such a simple matter to achieve the desired effect. As with all steam powered vehicles of the time, there was an inherent lack of expertise making it very difficult to produce an engine which would even move itself, let alone perform a task such as digging, as well.

It must be said that rotary diggers were in general poorly designed and in some cases badly constructed, two more factors militating against success.

Several engines relied on propulsion by means of the reaction against the soil from the rotating cultivator, a theory which would have been practical only if the soil was of completely even consistency. Any stones or obstructions meant a jerk forward of the engine with consequent strain on the cultivating blades and uneven wear on bearings, gears and chains, the latter of these frequently snapping. Of course the problem does not arise with a present day

A. Transverse bar carrying the diggers.
B. Links at end of transverse bars forming endless chains.
C. Rollers on short arm of transverse bars, which work over cams to give the necessary pitch to the diggers.
D. Chain-wheel fixed on main axis of machine.
E. Main axle forming fulcrum for bell-crank F and G.
H. Hand-wheel.
I. Counter-chain.
J. Spring seat for driver.
K. Steerage-wheel.

Fig. 5. Comstock's rotary spader. A horse drawn digger of 1873, power was transmitted to the tines from the wheels
(*Journal of the Royal Agricultural Society of England 1873*)
(*Cited J.R.A.S.E. hereafter*)
(*M.E.R.L.*)

rotary cultivator as it is so light that it can ride over obstacles, but when one remembers that a steam engine weighs perhaps 10 or more resistant tons, the difficulty of the steam pioneers can be realised.

The writer recalls using a small rotary cultivator on virgin soil surrounding a newly built house, and found that the machine would jerk when the rotating blades met bricks and rubble left in the ground by the builders. It was virtually impossible to control the machine and it was not until after removal of the rubbish by hand that any form of proper cultivation could be made. So if there was a problem in a domestic garden there must have been an even greater problem when cultivation of complete fields were anticipated with a primitive steam engine one hundred and thirty years ago.

The specifications of such engines as Holcrofts', Lillies and others similar, must call into question the engineering capabilities and qualifications of their designers who took an almost philosophical approach, hardly considering the practical aspects of putting their ideas into practice. They proposed the principle, made drawings, patented the idea and then left it to anyone who cared, to put the idea into practice if they could.

Of course the technology was not there to enable this to happen. Ideas, ambitions even dreams abounded, but when it came to the practical test, the lack of knowledge, materials, money and experience became all too apparent. This was not the inventors' fault, but was the result of circumstances in which they found themselves at that particular period in history.

Diggers of the latter part of the nineteenth century were not commercially successful for other reasons, for example the resistance to acceptance of new principles of cultivation, or lack of financial backing, but they were soundly engineered jobs, with adequate power available, and they were demonstrated at various shows and trials for all the agricultural world to see as (more or less) efficient, working machines capable of performing an efficient and economical task in agriculture.

The proposals of the 1850s however, did not have a hope of success from the moment any idea of steam digging was born, even assuming that archimedian screws and the like were not included.

Nevertheless, it was one of the more flamboyant characteristics of the Victorian age to produce such ideas, and one which makes the period so fascinating to look at historically, providing—if nothing else—a fleeting glimpse into a period the like of which will never be seen again. Some may say that is a good thing, and probably most would agree, socially speaking, but what a pity one of the prices of progress is a loss of individuality. There will probably never be another period of individual enterprise as seen in the nineteenth century and it is one function of the history of this period to remind us today, that we are still individuals, potentially as enterprising as any Victorian.

Rotary diggers were capable of fair work, but—particularly in the early days—too much was attempted at one time, the width of the rotary cultivator invariably being at least as wide, if not wider than the engine itself usually taking the form of a screw thread or paddle wheel, in either case perhaps more applicable to marine than to agricultural conditions. If they had been designed with cultivators of less width there would have been much more chance of success mechanically speaking, but of course the area of ground cultivated would be less for a given travel of the engine, making it unacceptable for reasons of running costs.

It was not until the introduction of a self moving engine with a second and independent drive to the rotary cultivator that some limited success was achieved. It is amazing that some of the rotary engines worked at all, and in the absence in many cases of any reports of field trials or tests of any sort, we can conclude that at least some proposed engines remained only as patents and ideas in the minds of the inventors, never reaching the constructional stage.

Chandos Wren Hoskyns was among the most voluble and vociferous exponents of the cause of rotary steam digging.

He not only wrote two books on the subject but backed his words with practical work in the form of a digging engine described in later pages of this book.

His words, entitled "Inquiry into the History of Agriculture" published in 1849 and "Talpa: or the Chronicle of a Clay Farm" published in 1853, are an almost fanatical recommendation of the finer points of rotary digging as opposed to ploughing, using the power of steam.

In his "Inquiry" he says:

"To those who are familiar with the achievements of steam power in so many of the most complicated mechanical processes, it must always seem surprising that its application to that of cultivation, though so often attempted, seems to have so generally failed. Looking back to the history of mechanical art, as it has been developed during the last half century, it would seem upon the whole a fair matter of surprise that it should not have been accomplished.

"It is worthwhile, however, to ask what is the specific impediment that forbids the banns between the Steam-engine and the Plough-share? What is it that prevents the versatility—(that peculiarly marked attribute of steam power) —which can drive a vessel of several thousand tons across the Atlantic, against a head-wind and sea, or spin the finest thread with a touch more delicate than the human thumb and finger—what prevents it from being applied to the clumsy performance of the Plough?

"Because it *is* a clumsy performance: and that noble power will have nothing to do with it. It is a law to which the annals of invention have given repeated proof, that late-discovered powers of Nature will not 'gear on' to those means and appliances which they have antiquated. They refuse to waste themselves. From the natural sympathy—so to speak—which exists between the 'best of its kind' in every department of matter, may be deduced the perception of a corresponding law of antipathy between things separate and incongruous in their nature and degrees of excellence, and remote in the order of invention. It is not the inapplicability of steam power, but the incongruity with it of the Plough (an instrument employed for the purpose of applying animal traction to the act of cultivation, and belonging only to that secondary class of powers), that forbids the union.

"The plough, and all the instruments that follow it, are only the 'animal-power' substitute for that more perfect process accompanied in brief by the spade, when worked by the foot, held by the hand, and directed by the skll, and purpose, of manual labour. This is what mechanical power must imitate: not the sluggish cleaving of the ploughshare, which only splits up an unbroken seam of surface, making a fulcrum of that which lies below, and thus pressing and polishing the sub-soil year after year into barren and impermeable induration, which the roots of no annual can penetrate. This is the first of a whole series of imperfect processes, not one of which it is even desirable to imitate; not one of which is necessary where the spade can be used.

"Why then should we wish that ploughing should ever be done by steam-power, stationary or locomotive? What we want is not ploughing, but cultivation: that process which the farmer by necessity performs in three, four, or five acts, not half so well as the gardener accomplishes it in one. As well might we expect to apply successfully the boiler and cylinder of a locomotive to the pole of a four-horse coach, or the shafts of a waggon, or the lever-handle of a common pump, or the distaff and spindle of a cottage spinning-wheel, as attempt to gear on the power of steam to the elaborate clumsiness of a plough.

"The gardener scarcely permits a dog to walk over a bed that has been newly worked; yet the farmer is obliged to let his whole team of horses with all his heavy implements pass over his land many times after the cultivation is finished; and even after the sowing is done, the seed harrows do but skim and film over the dismal work made in damp weather by the tread of the horses that draw them, and the previous implements: in fact it limits the cultivation of such soils to seven months out of twelve. Now all attempts at cultivation by steam seem to have failed, chiefly for this reason, that the experimentalist has set out with the idea of an instrument that is to be drawn backwards and forwards across the field, like a plough, either by a locomotive or a stationary engine. No such necessity exists. The spade is not drawn across the field: it acts perpendicularly upon the spot it is applied to; separating, lifting, and inverting each spadeful in succession, neither damaging by any farther pressure the soil it has once moved, nor hardening the subsoil underneath, in the act of moving it."

In "Talpa": "I say the Plough has sentence of death written upon it, *because it is essentially imperfect* (Fig. 6). What it does is little towards the work of cultivation; but that little is tainted by a radical imperfection—damage to the subsoil, which is pressed and hardened by the share, in an exact ratio with the weight of soil lifted, plus that of the force required to effect the cleavage, and the weight of this instrument itself. Were there no other reason for saying it than this, this alone would entitle the philosophic machinist to say, and see, that the plough was never meant to be immortal. The mere invention of the subsoiler is a standing commentary on the mischief done by the plough.

"But the willing giant stands idly panting and smoking: for nobody can agree to tell him what to do. One says, 'go and *plough*!' another says, 'go and

Fig. 6. "Steam power—no more to do with the plough than a horse has to do with a spade"
(Hoskyns "*Talpa*", 1853) (M.E.R.L.)

dig', each mistaking the means for the end, and trying to yoke this youngest born of human genius to the peddling routine of manual or equine capacity; out of the very perversity of backsightedness that clings to forms and modes which belonged to the *implements* not to the *task*—backsightedness that would with equal reason puzzle its brains in looking for the pole and splinter bar of a locomotive, the pendulum of a watch, or the paddle boxes of a screw steamer."

Quality of the work produced by later-made forked diggers was generally acknowledged to be of a very high standard, the one fault perhaps being that it was not deep enough. A depth of 6 to 8 inches was usual and was quickly seized upon by opponents of steam digging—who were many—their complaint being that the subsoil was undisturbed. Unfortunately the power available was insufficient to allow an increase of cultivated depth, although attempts were made in several ways to overcome this. For example, by adding tines to precede the forks and assist in breaking up the soil before the forks reached it. Spring loaded forks were also used.

A further potential hazard to forked diggers, as with rotary machines were stones in the soil which could cause a great deal of damage not only to the forks, but also the the driving mechanism and gears. Alleviating this may also be one of the purposes of spring forks previously mentioned.

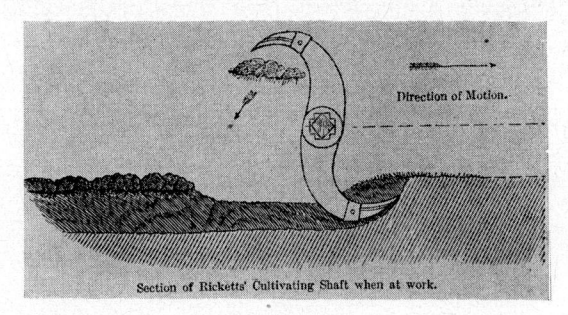

Fig. 7. **The principle of Rickett's rotary cultivator**

Both rotary (Fig. 7) and forked diggers were made that actually did work—even if it was with varying degrees of efficiency—and they could be found showing their paces at various Royal Agricultural Society shows and other public demonstrations although they did at times have the unfortunate habit of breaking down, usually for trivial reasons. No doubt this did not help their reputation.

Why did digging engines not achieve any success?

There were several reasons, not the least being the stubbornness of farmers in purchasing a new type of machine. Never before had they been offered a choice of method of cultivation. When the alternatives were placed before them they tended to remain loyal to the plough, rather than venture a great sum of money on a comparatively untried machine. The word did not get around that these new fangled monsters would do a better job than the plough, in spite of the fact that the few who did purchase or hire diggers usually became enthusiastic supporters of the idea, claiming economy of work and increased crop yields.

In spite of generous publicity the makers made little effort to advertise their machines. Most of the information contained in these pages is gleaned from show reports and a greater number of technical journals and newspaper reports. Only two catalogues exclusively devoted to diggers has come to the writer's attention. Patents have also provided information, although subject to their usual inaccuracies in certain cases are the only information available. So, were the manufacturers the ones who were lagging behind in selling their machines? Were they concentrating on perfecting their designs rather than marketing them? One has the feeling, when studying the whole subject of steam digging that the manufacturers rarely, if at all reached the stage where they were completely satisfied with the performance of their machines. Certainly Thomas Darby seemed incapable of standardising a design of digger, changing from broadside to tender attachment diggers, every illustration and description varying. This applies to most other machines reaching the stage where more than one was constructed.

It is difficult to estimate the total number of diggers of all makes and types that were built but the writer estimates 150 to 200. Compare this with tens of thousands of humble portable engines and many thousands of conventional steam ploughing engines and tackle.

Most forked diggers in their latter years of development—when perfected as far as they went—were exported to Europe and Egypt. The reason for this is one further reason for the failure of diggers in Great Britain. As has already been mentioned the quality of the soil was of great importance to efficient working of diggers. In addition to solid obstructions in the soil, it was also essential that the soil itself was not too soft as sinking of the digger could occur. The climate of the British Isles tends to make for soft conditions while the soils of Europe are somewhat firmer and gave the heavy machines a better chance of working efficiently.

At least three digging engines were produced in Europe, mostly in France (Fig. 8), but like cable steam ploughing, diggers in general and broadside diggers in particular were a peculiarly British and Victorian invention.

Another somewhat strange fact is, that to the best of the writer's knowledge, not one steam digging engine of any type whatsoever survives today. This seems even odder when one considers the inherent interest which strange machines stir in preservationists breasts. No one has seen fit to preserve one of these engines, but it is just possible that one day—if the hopes of the writer are fulfilled—that once again the forks of a broadside digger will fly. More of this in another chapter.

Perhaps there is a Cooper digger rusting away somewhere in Egypt. Should any reader know or even suspect the whereabouts of a steam digger of any make or type, the writer would be most interested to hear.

The most notorious of all Diggers were without doubt the Darby broad-

Fig. 8. A rotary cultivating engine as designed by B. P. Nubar. The engine is possibly of French make and constructed about 1890

siders (Fig. 9). Their notoriety far outweighed their numbers—about twenty-one machines in all. It was not until the introduction of tender diggers that at least some commercial success was realised, possibly because some of the well-established makers of road and ploughing engines—for example Fowler, Ransome, Sims and Jeffries and others—took the idea up and attempted, albeit halfheartedly, to make the idea of diggers a workable and saleable proposition. It was certainly not for want of effort on the designers part that diggers did not have any more impact on the agricultural world.

The universal drawback of the power/weight ratio already alluded to applied to both rotary and forked diggers, and was more of a problem with diggers in general than with other forms of mobile steam engines, due to the greater amount of machinery required to operate the cultivating mechanism which was always attached direct to the engine itself.

Many other types of steam cultivation engines were designed to keep the weight off the land being worked, due to the old trouble associated with horse ploughing, that of panning the soil. However, the nature of a digging engine determined that its whole weight was placed on the ground it was cultivating, thereby re-introducing the problem, even when the engine passed over before digging the soil.

Fig. 9. Re-coaling and oiling a Darby broadside digger. Tines flashing in the sunlight and with a full head of steam, this 8 H.P. Savage built engine was turning a deep spit of well tumbled soil.

There were some advocates of broadside digging. For example in 1886 G. C. Phillips, a partner of T. C. Darby said in a lecture: "The digger is virtually a steam man, with an iron arm and an iron foot, or rather a regiment of men, who exert their force with the least possible waste of power, and place all the advantages of spade husbandry within the farmer's reach at a very large reduction in cost. So that, practically, garden cultivation is available over all his farm. Its productive powers are, by this kind of tillage, developed in a surprising degree—a degree which will scarcely be credited by those who have not seen its marvellous effects." A small advantage over cable ploughing tackle was that a digger could cultivate the area being worked almost completely, including the headlands, but this did not compensate for the compacting and frequent bogging down.

These problems were never satisfactorily overcome during the age of steam, and it was not until the advent of the internal combustion engine that any improvement was shown with machines like the Gyro tiller, and present day small and light machines.

It was said that one advantage of the digging engine was its capability of being used for other jobs about the farm, but one is not able to see the benefits over the traction engine and in the case of the unwieldy broadsiders, a distinct disadvantage.

The steam digging engine was a machine fraught with difficulties in design, manufacture and use. Against the traditional plough and established ideas, it did not have much chance of success, although no doubt digging enthusiasts of the time would not have agreed.

It is not so much the story of the technical conquests of men, as one which shows the steadfast purpose with which some inventors pursued their ideas, yet in the end failing. Is their story any the less interesting because of this? Perhaps they deserve at least a portion of the glory that they would have won if they had succeeded.

Fig. 10. Broadsider made by Agricultural and General Engineering Co. of London
(*Essex Record Office*)
(*Cited E.R.O. hereafter*)

CHAPTER TWO

Ambulating Broadsiders

THE STRANGE SAGA of walking and wheeled broadside digging engines is one which hitherto has received scant attention. An historical omission which hopefully will be remedied in the following chapters. Many steam engine enthusiasts have heard of the mysterious Darby Broadside Digger in much the same way as ornithologists know the Great Auk—as some extinct but fascinating object about which only the Victorians were aware. It is with pleasure that the writer has been able to assemble enough information to relate the story of these ill fated mechanical curiosities.

Some present day enthusiasts labour under an almost romantic misconception that steam engines came into existence as if they had at some time been turned out of a factory in much the same way as cars are today. The misconception is complete when preserved engines are painted and polish applied to their massive ironwork, with little thought given to the manufacture of the same ironwork as long ago as 100 years. Although after 1900 the situation eased with the introduction of heavier and more efficient mechanisation, the sheer hard manual work involved in conceiving, developing and making a heavy steam engine—be it road, farm or rail—would quickly dispel any mass production ideas.

Possibly the most difficult path of all was taken by Thomas Churchman Darby of Pleshey Lodge Farm near Chelmsford, Essex, who in 1877 took upon himself the task of developing an idea which he had regarding a new method of mechanically cultivating the land by means of steam power.

Darby made himself responsible for every aspect of his project from financing and design through to manufacture and sales of this engine by the establishment of Companies.

Why did engines such as these exist at all? It may be thought with some justification that broadsiders were the product of the mind of a mechanical crank. Not so, Darby was a man with a firm idea of an alternative method of cultivation to the traditional plough. The first engine was constructed about one hundred years and the last one some eighty years ago. In these twenty years a total of some twenty machines were made all of similar design, and produced by several manufacturers. None of them now exist, but the evident tenacity of purpose behind Darby's work is clear.

He was born in 1841 and died in 1917. His roots were in Sheepcotes Farm at Little Waltham, Essex, where a series of John Darby's had farmed for two centuries, and his seventy six years passed in brilliant fashion, with a drive which should have borne greater success. It could be said that he was the last of the independent engineering farmers in the Mechi tradition.

At the age of seventeen he patented a horse hoe which proved to be a most useful and efficient implement, being awarded a Medal at the Essex show of 1863. Some of the hoes were still in use some eighty years later. By 1862 a small iron works and smithy was established, which produced the hoe and other agricultural machinery.

While still at Little Waltham, first thoughts regarding his steam digger were taking shape. Soon after production of the horse hoe ceased, the family moved to Pleshey. It was here that ideas crystallised into drawings and drawings into metal.

The principle of Darby's machines was based on the premise that dug as opposed to ploughed soil provided a better tilth, affording seeds a better chance of germination and weeds less chance of survival when completely dug in. This was acknowledged as a fact in the agricultural world, but it was also recognised that to dig by hand was very expensive in terms of manual labour. Under ideal circumstances where complete spit depth was attained, it was superior to ploughed land, but it was a common fault that labourers took every opportunity to "scamp" the work, leaving it of beautiful appearance but digging to a shallow depth not at all suitable as a seed bed, a complaint which John Knight with his hop digger, also voiced.

Thus digging cultivation obtained a bad reputation, with its expense and inefficiency, which Darby sought to correct by the application of steam power to a many forked engine which would not skim over the surface in human fashion, but dig unfailingly to full spit depth at an area rate comparable with the steam ploughing set of tackle with which he was in direct competition.

Darby was starting his enterprise at a time when the steam plough was already at an advanced stage of development—an additional commercial hazard for him, but he proceeded undaunted.

What exactly was a broadside digger?

It was an engine which broke away completely from traditional design, treating the problem of cultivation in a new way, being designed to suit the basic idea of powering a fork.

The concept was revolutionary—probably too much so for the conservative farmer. Ideas such as Darbys were so different that most of the traditionalists took one look and fled in horror from such an odd looking monster!

The boiler—with the exception of the first two or three engines—was of a standard design on all broadsiders. It was so designed that the firebox was situated in the middle of the boiler length, giving effectively two boilers attached end to end, with a common firebox, two barrels with two sets of tubes and two smokeboxes and chimneys, this last feature being unknown on any other type of agricultural engine boiler, although boilers with central fireboxes and a single central chimney, have been made and used on other road engines—for example on the Yorkshire steam wagon, and some of Howards ploughing engines. The "Fairlie" locomotive utilised the same boiler as Darby, and one is still used on the Festiniog railway.

Along the length of and to one side of the boiler were arranged a series of forks which were operated by a system of shafts and bevel gears. These were driven from the crankshaft of the cylinder and motion mounted on top of the boiler, the cylinder being at one end and the crankshaft the other.

Wheels, also driven by bevels, were arranged in two pairs at each end of the boiler and were so mounted that they could be swivelled through 90° thus enabling the engine to travel "lengthways" along the road or from job to job. In the other wheel setting the engine when working could travel literally sideways or "broadside" on to the land it was digging. When at work, an outrigger was attached on the opposite side of the boiler to the forks serving the purpose of providing steerage to the engine while travelling broadside on. The outrigger was detached when the engine was converted to road setting, and towed behind.

Two speeds were provided, one for road travel and a greatly reduced speed for digging, sometimes by means of a set of gears working on the principal of a lathe back gear.

The only broadsiders to vary from this general design were the first two and last engines constructed, having a standard type traction engine boiler on Darbys first two walking diggers and non swivelling wheels on the small 6 H.P. machine made by Savages in 1888 which—as it were—travelled permanently sideways. But more of these in the following pages.

Darby spent years in considering alternative methods of steam cultivation.

His personal experiences gained while farming convinced him that the plough was out dated. "I felt that what was wanted was direct action upon the soil, in contradistinction to the wasteful practice of dragging the plough through it", he wrote.

Armed with the completed drawings of his digger, but with a complete absence of practical digging experience, he tried to persuade at least one firm to construct it. "It was in 1877 that I took my drawings to Messrs. ——. They would not take it up because they were of the opinion that I should never get the machine across the field."

Which unenterprising firm this was is in doubt. The names of Coleman and Moreton, and W. & S. Eddington have been suggested, but it is unlikely that it was the latter of these as Sylvanus Eddington was very much involved with Darby and his diggers almost from the beginning although it is possible that help was given unofficially by Sylvanus, using the Companies workshops in Chelmsford.

An undiscouraged Darby proceeded to build his own workshops at Pleshey specifically for the purpose of constructing diggers. Time was limited and work proceeded apace. A number of fitters and a foreman were engaged to start work on the prototype engine. A major piece of equipment was a lathe, powered by horse gear and later by steam. Sylvanus Eddington produced the castings in Chelmsford, which were taken to Pleshey, machined where

Fig. 11. Thomas Darby's walking digger. This incredible engine was either the first one or the second one rebuilt from the first. Shown during trials at Pleshey about 1877

necessary and assembled. As the sheds clanged to the sound of the hammer on rivet, the revolutionary shape of the first walking broadsider took shape.

One of Darby's sons, Sidney, started at an early age assisting his father with drawings. He was asthmatic and did not go to school until he was thirteen, and spent some of the time taking drawings of foundry patterns to the village coffin maker, where the wooden patterns would be made for use at Eddingtons foundry. "I didn't like it when someone died" reminisced Sidney many years later; "that held the work up so."

The first broadsider—an amazing pioneer engine—(Fig. 11) supported its 9 tons entirely on six legs, which through a series of levers and cranks, walked the machine along on plate steel feet. Seeing this machine in the field must have been an incredible sight. One which quickly convinced its designer that it was a case of "back to the drawing board". It rocked, heaved and swayed as it plodded along, no doubt giving the driver a frightening time, being prepared to jump for his life lest the engine toppled over. A single 8 H.P. cylinder atop a long barrelled, but otherwise standard type of traction engine boiler, provided the power for both "ambulating" and digging. The idea of the need for turning the soil completely over, had persuaded Darby to design the fork action to incorporate a twist to one side of each fork after plunging into the soil and

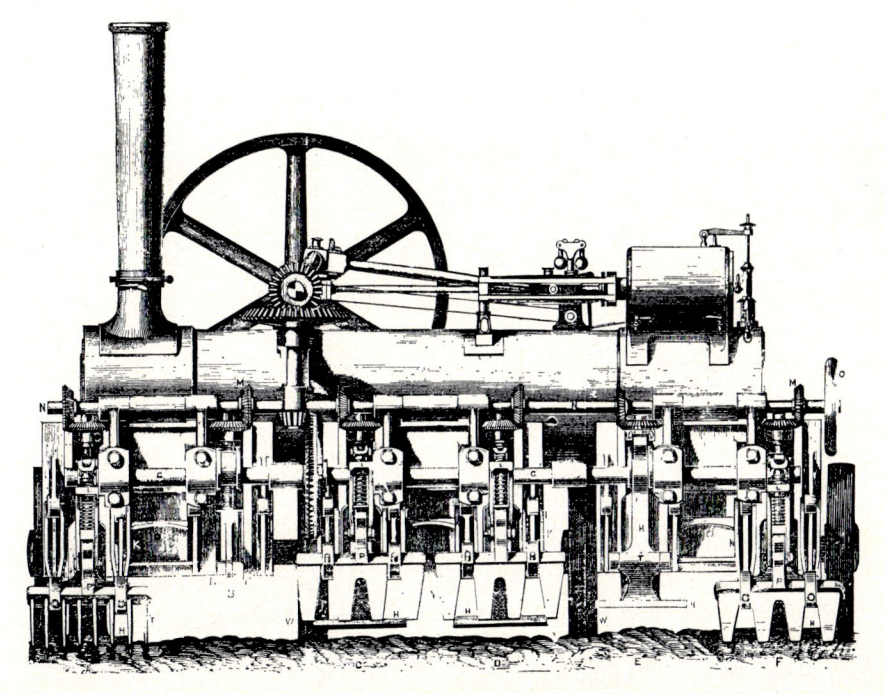

Fig. 12. This modified broadsider retained the ambulating mechanism of the first engine, but wheels now aided stability
(*The Engineer, 1878*)
(*M.E.R.L.*)

Fig. 13. The stabilising wheels turn through 90 degrees, enabling the engine to move lengthways along the road, with forks folded up
(*M.E.R.L.*)

lifting it up, imitating as closely as possible the action of a human while digging. The mechanical difficulties of executing this action to six pairs of forks are enormous. It is a credit to Darby that they worked at all, albeit inefficiently as trials proved. He eliminated the twisting action on all subsequent machines, relying on tumbling of the soil to invert it, and concentrating on penetration to as great a depth as possible in order to break the hard pan. This feature was considered more desirable. It was a major criticism by many (perhaps unfounded) that the broadsider compacted the soil into a worse condition than before, forgetting that the machine performed its work *after* passing over the soil.

The first digger had a very short existence, for when it was worked for the first time, it plodded round the field once, digging as it went. This was enough for Darby who said: "I saw it wouldn't do; it jumped too much; there was too much of the goosestep about it; it was very ridiculous".

"I knocked it all to pieces and made it all over again," says Mr. Darby; "this time I made it with broad wheels in its rear, and feet in front, for I was still somewhat attached to the idea of the pedestrian movement (Figs. 12, 13, 14). With this machine I dug 70 or 80 acres of satisfactory work right off, and everybody who saw it thought it was a wonderful thing and a perfect success. The parts of both these pedestrian machines were made at Chelmsford, and I put them together by my own men in my own sheds at Pleshey. The next machine was made entirely at Chelmsford, but was too heavy to be of any use; it cost £1,700, and never did any work on the land. I have it now in the workshop, and use it for driving lathes and drilling machines."

"That looked as if it wer' a-gooin' along on stilts" one old man remembering seeing the walking digger reminisced in later years.

Horses shied at the first engine, its ponderous parts working along amidst clouds of steam. Farm workers observing the monstrous mechanical giant said that the beat of the engine straining to make its thick tined forks penetrate the soil, spoke in rhythm, hissing "I hate this work, I hate this work". One reason for the unpopularity of the engine was that it could take as long as a day merely to shift it from one field to another.

"Still enamoured of the pedestrian principle, I again decided to make a machine which was semi-ambulatory, but that should be lighter. This was constructed partly at Chelmsford and partly at Leeds; it worked well the whole season, dug land for neighbouring farmers, as well as my own farm, and gave

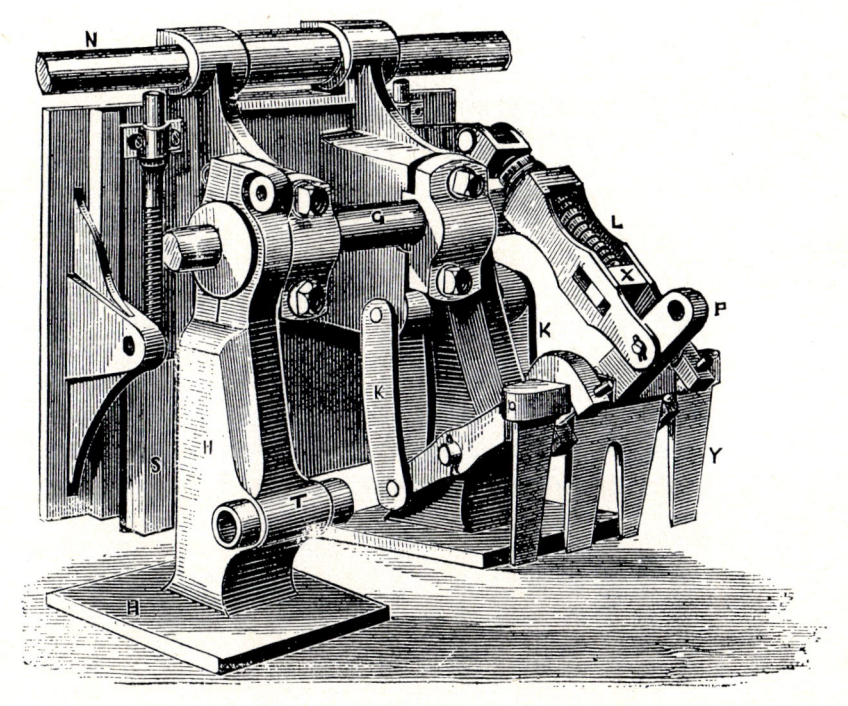

Fig. 14. Mechanism of the "feet" and forks of the 1878 walking digger
(*M.E.R.L.*)

THE PEDESTRIAN BROADSIDE DIGGER.

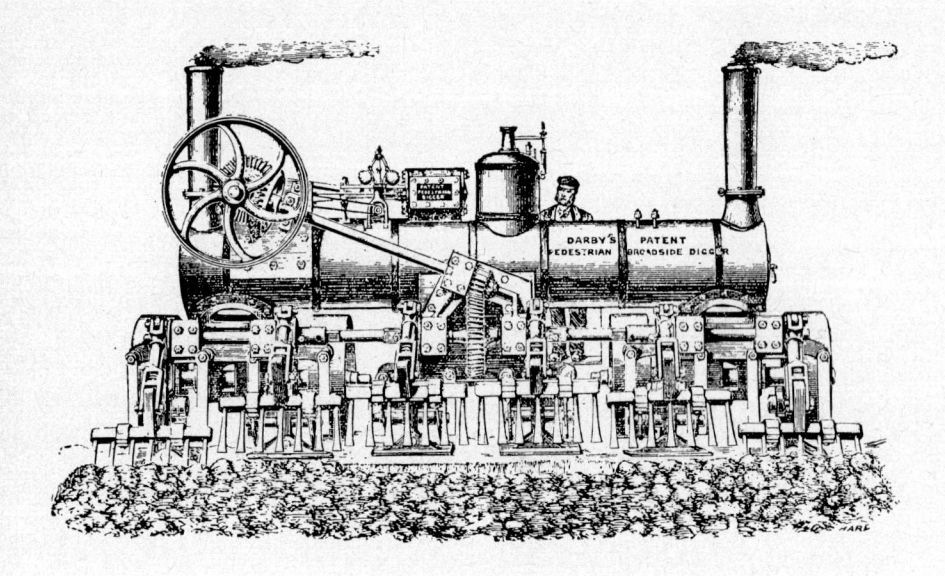

Protected by the Patentee in England, the United States, Canada, Germany, Austria, India, France, and Belgium.

Fig. 15. Darby's Pedestrian Digger in its final form as shown at Kilburn in 1879. Subsequent engines were completely wheeled
(M.E.R.L.)

much satisfaction (Fig. 15). This was in 1879, and the machine was exhibited at the Royal Agricultural Society's Show, at Kilburn, in that year."

This extract from Darbys own description of his workings, shows that four walking diggers were made in all—three by Eddington & Darby and one by Eddington & McLaren (the Leeds firm) indicating in this one and only reference the early involvement of McLarens with Darby's broadsiders.

Darby had not finalised his ideas on the mechanism of his diggers before the prototype was made. He obtained a provisional patent on 4th June 1877. It was voided "by reason of the Patentee having neglected to file a specification in persuance of the conditions of the Letters Patent". It is interesting that this first digger patent described an engine supported on three rollers, two of which were driven by worm gears to move the engine along while digging and the third one set some distance behind to provide steerage. He went on to specify a device which did not appear on any broadsiders, but could have been useful. "Projecting from the rollers are circular discs or knives; these are carried at intervals apart upon the rollers corresponding to the distance between the digging tools, so that as the implement travels over the land the knives cut the land into strips of equal width, and each of such separate strips is afterwards acted upon by one of the digging tools; the digging tools are, as before stated, ranged in a row along the side which now forms the rear of the implement." Effectively, he is saying that some form of prior stirring or cutting of the soil should be made before the forks actually dug the soil. This is an idea re-thought of by Savages of Kings Lynn in later years. The surviving works drawings of their digger shows several versions of fork design. One of them shows a coulter preceding the fork, although it was not actually used, simplicity being preferred. It was fairly evident that the first specification would have made a better road roller than digger with the land compacting severely—so it was allowed to lapse.

It was not long ere a more practical specification was lodged with the Patent Office (Fig. 16). In July 1877—at about the same time as the prototype engine was completed, he described a new engine which could either be supported on driven rollers—as previously—or supported and propelled by mechanically driven feet and legs. The attached drawing shows an engine remarkably similar to the prototype shown at Pleshey and one can only assume that the rollers were left on the patent to confuse would be imitators as to his intentions.

The third walking digger was made between 1879–80 at Eddingtons workshops.

Although mentioned by Darby in his interviews it seems that he considered it a complete failure. No illustration has survived.

Trials proved to be a disaster. Cast iron components cracked under the strain leading to the writing off of the engine as a mechanical failure, and it was relegated to the humble task of powering the workshop lathe, working for many years. Later, advertisements proclaimed that all working parts were "Made of the Best Steel".

A patent was lodged for the design of the Fourth engine in May 1879, generally specifying it (Fig. 17). Darbys agile mind even at this stage was thinking of other methods of digging and the same patent illustrates three alternative designs of engines. One is very similar in layout and operation to Barrats "la piocheuse" engine, the only difference being the inclusion of a rotary movement of the "pick axes" or mattocks rather than a reciprocating one. Another design shows an engine with an outrigger, similar in fashion to those incorporated on later broadsiders. The third design is of a rather ugly carriage, the boiler on top and the digging forks below. On large wheels, the carriage frame was far too high and the whole design unworthy of Darbys technical brilliance. However the outrigger had now been thought of and it was included on the third broadsider.

Spade husbandry was claimed to be universally acknowledged to be more effectual than other systems and Darby was not slow in pointing out the potential advantage of his digger for this purpose.

Better aeration and finer tilth were claimed for the soil and—a touchy point with Show judges as William Smith of Woolston had discovered—by adjustment of the fork operating mechanism, the soil could be broken up without being turned up on edge suitable for fallow, stressing the point that the "earth can also be *turned completely over* as in ploughing and left either in

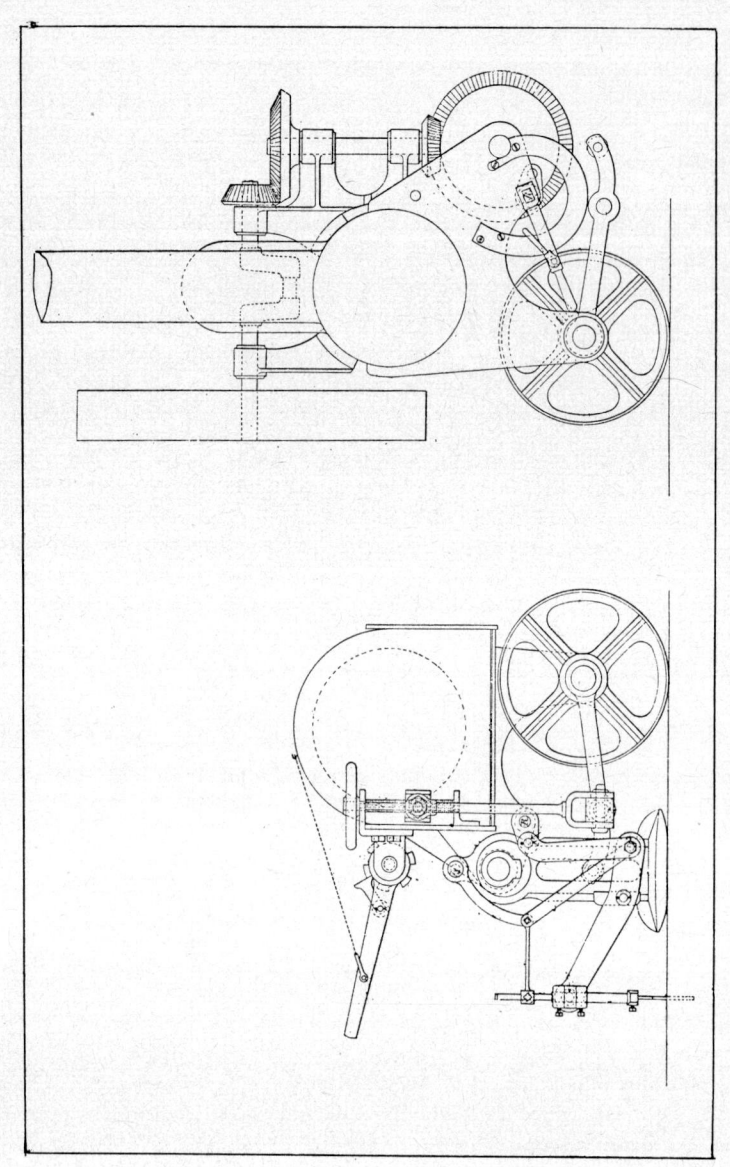

Fig. 16. Darby's specification of 1877 illustrates the combined wheeled and walking motion of the engine
(Patent Office)

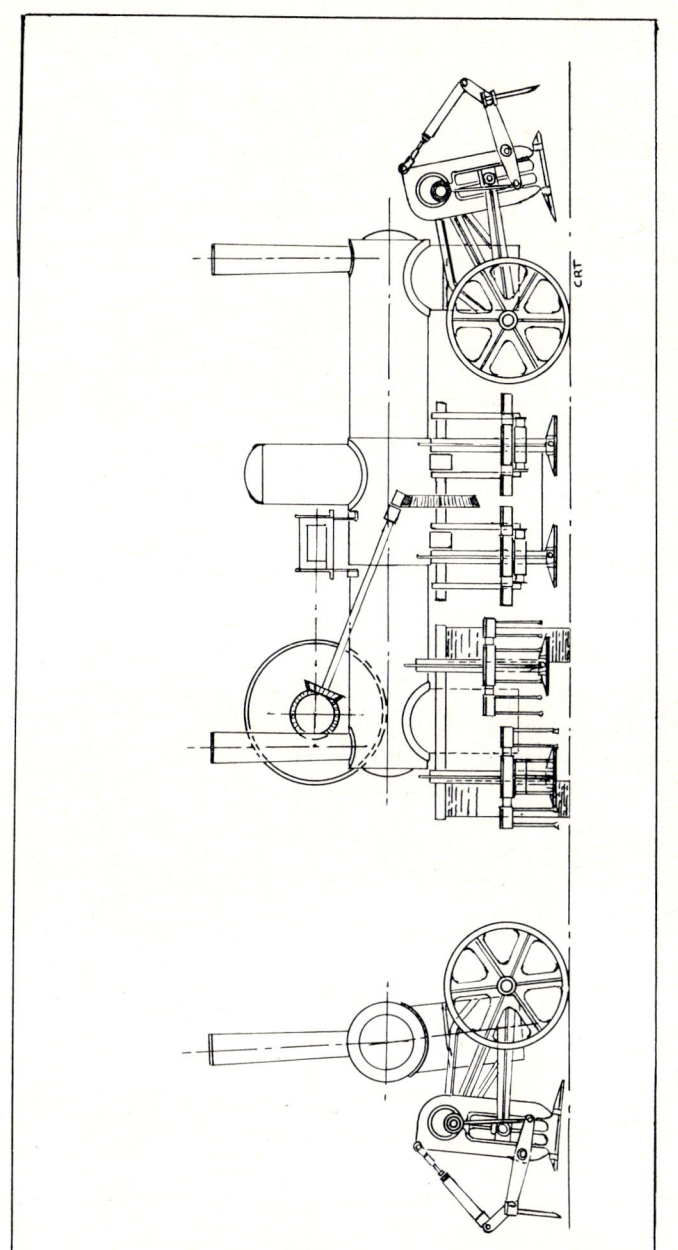

Fig. 17. A later specification of 1879 shows an engine which bears strong resemblance to the fourth engine exhibited at Kilburn
(*Patent Office*)

large pieces, or broken up small; in any case the subsoil is always moved deeper than that which is actually turned up".

Irregular shaped fields, the short lands and headlands could all be dug with ease simply by driving over the ground, albeit sideways and this was an advantage over steam ploughing.

Low capital and running costs, minimal labour and the ability of the digger to be turned to other uses—threshing and grinding for example—and (shades of shift work!) the digger could even be adapted to generate its own electric light for round the clock operation!—were all claimed for the broadsider.

Relative operating costs between digger, steam plough and horse plough were claimed to be £2·35, £4·75 and £6·00 respectively per day on ten acres, although the digger amount must have been estimated as there was little practical experience behind it at that stage. Engine prices were only to be obtained on application and were probably never accurately fixed as in the event the walking diggers were found to be wanting in their performance, the main snag—apart from snapping castings—being the erratic motion imparted to the diggers by their legs.

At this point, a description of the factory and works where most of the prototype and development work on broadsiders went on would be appropriate.

W. and S. Eddington of New Street Ironworks, Chelmsford, were a small concern run by William and Sylvanus. They produced their first traction engine in 1880 and went on to produce a series of well designed and built engines with both single and tandem compound cylinders. In 1886 a Mr. Steevenson joined the Company when the name became Eddington and Steevenson. The partnership was dissolved in 1890 when Steevenson joined Darby in producing the Darby-Steevenson digger, described later. The ironworks and its contents were sold by auction in October 1894. If broadside diggers had "caught on" it may have been a different story.

In 1877 Sylvanus Eddington was making castings and machining the larger ones for Darby's first two walking diggers, subsequently making and assembling the third (unsuccessful) one entirely in the New Street Ironworks. The exact proportion of Darby's work executed by Eddingtons is unclear. Parts moved freely between Darby's sheds-cum-workshops at Pleshey, and Chelmsford. Diggers were assembled at both works, although the greater number at Pleshey. Sylvanus was responsible for supplying Darby's needs, little being known of William's connection with the digger project. One wonders if he was the reason for Eddingtons being suggested as the firm who turned down Darby's proposals in the early days.

Darby's workshops were no more than a collection of sheds (Fig. 18). The largest of these housed the engines currently being assembled, while the illustration shows several diggers in varying stages of dereliction or completion laying about the farm, a portable engine in amongst them and chickens no doubt roosting in the boilers. Drawn about 1886, the sketch shows a busy yard with one engine half in and half out of a shed. This is the third walking digger which was scrapped, and is now powering the workshop lathe. The scene has a delightfully rural atmosphere about it, rarely seen today. Interestingly, the complete engine, in the centre of the picture, appears to be of the same design (if not the same engine) as the one in the advertisement of the Agricultural and General Engineering Co.'s of 1882, described later.

Fig. 18. Darby's country workshops sketched at a busy period
(E.R.O.)

"A very curious, not to say funny invention attracted daily crowds of gazers. The irreverant styled it the 'wary wobbler': its being, end and aim was to irrigate, and as it wobbled over the grass, distributing spray on every side it seemed to possess not merely instinct or volition, but even reason. ... In turning, one of the inner legs marks time, while the others wheel round with a precision unsurpassed by a regiment of Grenadiers."

Under review in the *European Mail* in 1879 was Darby's fourth Pedestrian Digger, as it was dubbed.

Exhibited in 1878 at the Royal Agricultural Show, this engine had been preceded the year before by the appearance of one or possibly two scale models, shown at Bristol and Islington Cattle Show. The fate of these models is unknown, and should any reader be aware of their existence the writer would be most interested to hear.

At the Kilburn exhibition on stand 549 in front of the entrance to the Agricultural Hall offices and restaurant, the machine—shown as a static exhibit—gathered crowds whose reactions varied from fascination to ridicule.

A descriptive pamphlet, available to all interested, pronouncing the existence of the Darby Patent Pedestrian Broadside Digger Company Limited, the cover of which is the basis of the dust cover of this book, is also illustrated in its original form in Fig. 19. This is the only catalogue of its kind to have come to the attention of the writer.

The fourth walking digger made by Eddingtons and Mc. Larens according to Darby's catalogue consisted "of a 10 H.P. engine having its fire box in the centre of boiler; it is intended when digging to travel over the land broadside on, taking a breadth of 19 ft. 6 in. each journey. It is supported on one side by four wheels, each having a width of 12 in., the opposite side is supported on six feet which assist the locomotion when it is at work and from which it derives its name of Pedestrian. On this side of the machine are attached bearings carrying a horizontal shaft extending the whole length of the engine; upon this shaft

Fig. 19. The Kilburn Exhibition catalogue cover
(*M.E.R.L.*)

are six eccentrics by means of which motion is conveyed to the legs and feet working alternately by pairs, to the legs are attached six forks by means of which the digging is effected, these together with the wheels and feet take the weight of the machine evenly and equally. The coal bunkers and water tanks, foot plate, the fire door, and steerage wheel (which can be worked by the engine driver whilst standing in his place at the fire box) are fitted on the other side of the engine.

"The digging shaft is made in three lengths capable of being connected when at work digging and disconnected when travelling from field to field, the middle section of the shaft (with two legs and forks) is fixed to the back of the fire box and by moving the eccentrics to their highest position the feet and forks are raised clear off the ground.

"At each end of the boiler are fixed large studs, round each of which works a frame carrying on one side, the digger shaft with two legs, and on the other side two travelling wheels.

"For travelling from field to field these frames are turned a quarter round, bringing the wheels into proper position, by a simple arrangement the act of turning raises the feet and forks clear off the ground.

"In turning at the headlands one of the outside legs mark time, and the others step round like soldiers wheeling.

"The Digger only requires the attendance of one man, by whom all its movements can be easily directed; when at work it travels over the land at the rate of about half-a-mile per hour leaving the cultivated land in its rear. By having the firebox in the centre of the boiler the inconvenience of getting the fire box uncovered and melting the plug when on hilly ground is entirely obviated.

"In the manufacture of the Digger the very best materials and workmanship only are used. The boiler is constructed entirely of steel and double rivetted; the gearing, brackets, and all the principal parts of the digging machinery, travelling wheels, etc., are of best crucible cast steel; all shafts and axles of steel; the bearings are all made of a special gun-metal mixture which will wear much better than generally used.

"The working parts of the machine are so arranged that the greater part of the power is applied to the tillage of the soil close to the point where it is generated and is expended to the best advantage (like a man in the act of rowing), thus securing the benefit of steam cultivation with an engine of lower power, working at less cost than is possible by any other known system."

As previously mentioned, the first and second engines possessed a more or less normal pattern traction engine boiler, while the engine illustrated and described above from the Kilburn show catalogue, had the double ended boiler which became a universal feature on all broadside diggers. It still walked, but the forks were now divided into three sets. The middle set was driven as before, but the two outer sets were now mounted on pivoted axles each with a pair of wheels which could be turned through 90°.

About a month before the opening of the Kilburn show a practical demonstration of the digger was given at Pleshey Lodge and judging from the comments which followed, the machine must have performed well. Some remarks suggest that it was an earlier "rough style" digger, but this is not certain. It would certainly have saved cleaning up and rushing the engine to London for the show.

It was the fourth engine, with the refinement of the double ended boiler and improved finish which was shown at Kilburn.

A total of three walking diggers were made by Darby with some parts supplied by Eddingtons between 1877 and 1879, and one by Eddingtons and McLarens. The fifth one discarded ambulatory movement entirely and used wheels only to support it.

A soiree held at the Essex and Chelmsford Museum on the evening of 22nd January 1879 would seem to be an unlikely occasion at which to find anything connected with steam digging. Reading through the programme, however, reveals that the "Grand Scientific and Musical Soiree" of the Philosophical and Natural History Society was—despite the unlikely combination of ologies and organs—in fact an event well endowed with the latest wonders of the age.

On show were examples of the phonograph, telephone, stereoscope, a new organ and a model of Darby's Pedestrian Digger. Adding to the evenings delights, interspersed between short lectures, a Madame Mudie-Bolingbroke rendered her vocalisation of Victorian melodies to the visitors.

The ladies and gentlemen thronged the Shire Hall on the chill winters evening viewing the technical marvels on show. Amid the chatter, clink of tea cups, hiss of gaslight and Madame Mudie-Bolingbroke, the beginnings of many of todays technologies provided amusing entertainment.

Thomas Darby himself was present and was no doubt full of high hopes for the fulfillment of his endeavours when he demonstrated his model digger, which the programme optimistically stated was "intended to supersede steam ploughing". In between Mr. Hicks serving Tea and Coffee (2d. a cup) and the Electrifying Machine (shocks, gratis) Darby patiently explained the workings of his model to the sometimes curious and sometimes interested visitor, who could have gone away at the end of the evening with some strange predictions of what the future might hold in store. Darby had no such forebodings, and continued developing the steam digger.

CHAPTER THREE

Vehicular Broadsiders

AN EXAMPLE OF THE TRICKS OF FATE with which information sometimes survives was demonstrated when in May 1951 a photograph appeared in *The Model Engineer*. It was of an unknown type of engine, and sent by a reader who described how the photograph came to light.

"I picked it up by the roadside while out for a cycle ride; it had a name and date on the back and was in a leatherette folding frame facing a picture of *Victory* (Nelson's ship). The name was Plesbey, the date 31st May 1891."

It was of course a broadside digger at Pleshey and the photograph is reproduced in Fig. 20. A lucky event which gives a valuable bonus to the known details of the elusive broadsiders.

This engine is one of the first wheeled diggers made by Darby and Eddingtons. It is in a somewhat derelict condition, but the photograph provides a wealth of information which otherwise would have been lost.

Fig. 20. A derelict broadsider at Pleshey. Note the seat on the right hand side of the outrigger and the steering yoke often used on the larger diggers
(*Model Engineer*)

49

50   DIGGING BY STEAM

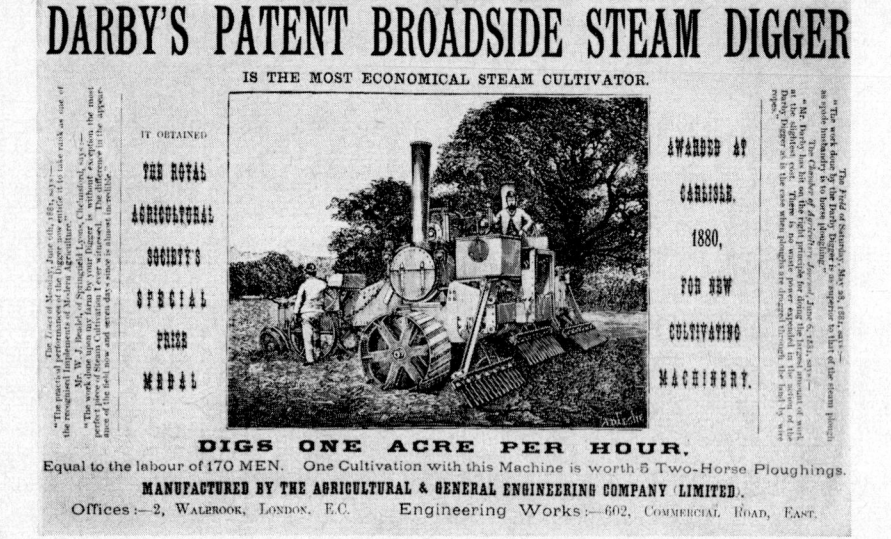

Fig. 21. Agricultural and General Engineering Co's advertisement of 1881. The outrigger is of somewhat different arrangement to other makes of broadsiders
(E.R.O.)

It is evidently a "steel" engine as flat plate bearers are used for bearing housings throughout in place of cast iron.

Notably an outrigger is incorporated—as on the third patent of 1879 and the pedestrian mechanism is discarded in favour of wheels.

Although taken some twelve years after its construction, the photograph reveals some extraordinary constructional features. For example the left hand chimney is 20% longer than the other. The left hand smokebox too, is longer than the other and has an angled door to clear the fork mechanism when turned. Could it be that two odd boiler barrels were utilised for economies sake? The forks have been removed and it seems likely that the engine was retired from working life.

Kelly's Directory in 1882 contained an advertisement (Fig. 21) by the Agricultural and General Engineering Co. Ltd.—incidentally, not connected with the firm of the same name, ill founded in later years as a combine of well known steam engine builders to ward off the threat of the internal combusion engine—with offices at 2 Walbrook, London and works at 602 Commercial Rd. East. Their success or failure in constructing broadsiders has not been recorded. Some distinctive features are shown in the illustration. Among these are extra wheels on the outrigger and angled land steerage wheels. A broadsider made by the Agricultural and General Engineers, was exhibited at Smithfield in 1881. This engine was reported as having five sets of forks and a cultivating width of 21 ft., while the advertised engine possesses only three forks (Fig. 22), perhaps indicating that more than one was made in London. Alternatively, Darby could have loaned them the illustration for the advert, as

Fig. 22. Agricultural and General Engineers broadsider which is similar to those in Figs. 10 & 23 in having three broad forks
(*Implement and Machinery Review 1881–2*)
(*M.E.R.L.*)

the same one appears quite often in various contemporary articles; and an engine of apparently the same design appears in the drawing of Darbys workshop yard.

Between 1880 and 1883 J. & H. McLaren undertook construction of Darby diggers (Fig. 23). The McLaren version incorporating swivelling wheels, similarly to the third Eddington engine. Four engines were made to Darby's new specification, one being shown at Carlisle R.A.S.E. in 1880.

Four McLaren broadsiders were constructed No. 82 was the first one and the one shown at Carlisle.

The other three engines were purchased by Wimshurst, Hollick and Co. of London. These engines were numbered 133 and 134 supplied in 1881 and number 135 in 1883.

The Carlisle report said "Of all the exhibits D D No. 3566 made by J. & H. McLaren of Leeds probably caused the greatest amount of interest: was by far the best response the Society had to its offer of trials for any new implements for the cultivation of the soil by 'steam or other mechanical force' and well deserved its Silver Medal".

Of 8 H.P. the engine had a single cylinder 9 in. dia. by 12 in. stroke. Two bevels were on the crankshaft one of which drove the road wheels at two speeds and the other driving the digger mechanism. This was similar to the third

Fig. 23. Probably the only engine to possess forks on the outrigger side of the boiler, this modified broadsider possessed worm and chain steering, traction engine style. Made by J. & H. McLaren
(*M.E.R.L.*)

Darby engine in that there were three sets of forks, the outer one having 14 tines, the central one 13, giving a total cultivated width of 20 ft. 8 in. The McLaren forks were situated on the opposite side of the boiler to other broadsiders, that is on the same side as the outrigger. Although not mentioned in the report, it would seem that the great width of the forks would have put tremendous strain on the driving gears and bearings, as with each revolution of the digging crank the bearings would have suffered three heavy thrusts of the forks, wearing quickly and unevenly. Other manufacturers in general halved the width and doubled the number of forks giving smoother wear to the mechanism.

The digging crankshaft drove the forks through Hookes joints, revolving once for 3·92 revolutions of the engine crankshaft.

Drive to the two pairs of wheels could be selected at will either all at the same time or singly. The outer ones—in the digging setting—could be disconnected by removing pins to facilitate turning at the headlands and the inner ones by clutches, thus giving suitable drive under all conditions.

Steering while digging was by means of the steel outrigger which was an essential part of all engines of this type after the "walkers". The outrigger served several purposes not the least of which was stopping the engine—with wheels aligned along the boiler in readiness for digging—from falling over! It also served as tender, water tank, and footplate as well as providing steerage. Possibly the essential siting of the outrigger on the engine was indicated, or at least influenced the design of the double ended boiler for which Darby diggers were famous. The central fire box was easily stoked from the outrigger tender, and also gave about the best available view of the diggings for the driver. At the other end of the outrigger, which was pivoted where it joined to the boiler, was mounted disc wheels—on McLarens engine 8 disc coulters—which bit into the soil providing steerage when they were turned by means of a handwheel on the footplate.

When the engine was required for road travel the wheels, which were 3 ft. 6 in. dia. by 2 ft. wide were turned through 90° and the outrigger detached from the boiler (Fig. 24).

At the Carlisle trials, McLarens broadside digger worked well in spite of two inches of recent rain which had turned the soil to mud. The show report stated that the work performed "was equal to and in some respects better than ordinary ploughing, but not so good as cultivating or digging as far as laying up the land rough for fallowing, which is generally wanted where steam is used, but could doubtless be altered". The engine was worked on a slope which in places exceeded 1 in 10, with the engine working at 200 r.p.m., it also dug on the flat and across some fallows, the last with difficulty in the wet conditions.

Some details of the McLaren broadside digger are:—

Weight of engine complete 15 tons 7 cwt. 3 qrs.

Weight of outrigger and disc wheels 1 ton 17 cwt. 1 qr.

Total heating surface (firebox & tubes) 171·3 sq. ft.

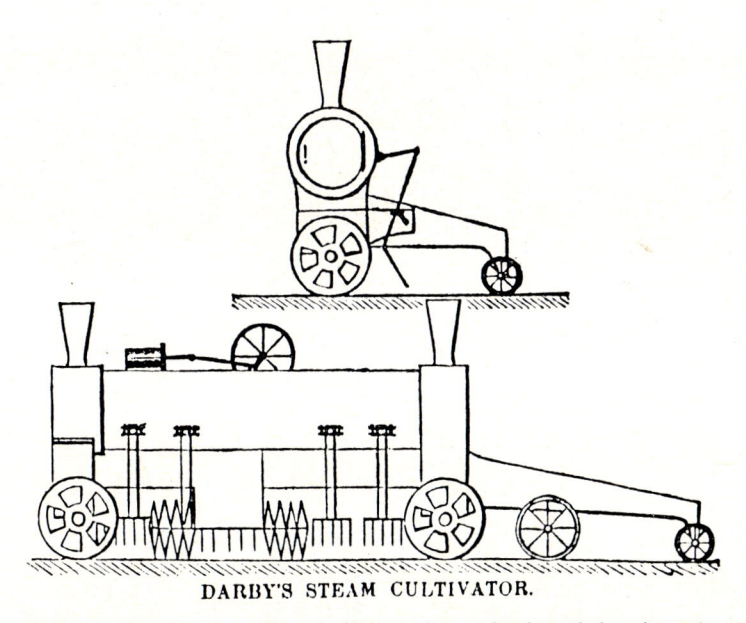

DARBY'S STEAM CULTIVATOR.

Fig. 24. This outline sketch is the only illustration so far found showing a broadsider in road travel setting. This is probably because most diggers spent a large part of their life on one farm, rarely venturing on to the open road
(M.E.R.L.)

Grate area 6·64 sq. ft.
Fork tines 6 in. apart.
Average depth dug 6·07 in.
Average R.P.M. 180
Area dug per hour (200 yd. furrow) 1·19 acres
Average indicated H.P. 28·3
Boiler working pressure 120 lb/sq. in.

That development of steam ploughing had literally become "stuck in a rut" was remarked upon in 1885 by *The Engineer* when reviewing the R.A.S.E. show at Preston.

"Steam ploughing machinery seems to have got into a groove, like almost every other kind of agricultural implement. The balance plough shows no desire on the part of the makers to offend by innovation." While diggers had not as yet reached a stage of perfection, at least lack of innovation was not one of the criticisms which could be levelled at them.

In contrast to the rather subdued reports of the R.A.S.E., *The Engineer* was both complimentary and critical of the McLaren built Darby broadside digger at Carlisle.

"The machine is admirably ingenious. Beyond any question Mr. Darby has solved a problem which has for many years vexed the souls of agricultural engineers, and we will go so far as to admit that he has sent to Carlisle, with the aid of Messrs. McLaren, a perfect digging machine—the first ever made. By some however, it is considered that it has yet to be proved that digging is better than ploughing. This is a point for the practical farmer to decide; and it must be proved that it is not only better, but very much better, for the objections to digging are very great. An engine weighing at least 16 tons has to be moved over the field; save on comparatively even ground places are skipped which would sometimes need to be dug by hand. This is lost in turning round, and to use the machine with economy the bouts must be long. Granting that the machine can get over the ground in any weather, the fact would remain that four rollers, each 2 ft. wide, and each loaded with 4 tons, must pass over the land for every 20 ft. of width dug. That these rollers will consolidate the land several inches cannot be doubted, and a repetition of the process a few times would, it seems to us, produce a hard pan, injurious to the growth of crops; but as the digging is done after the engine has passed, this objection may have less weight. Mr. Darby has, however, some difficulties to overcome in order to compete with the ordinary engine and wire rope. For instance it takes roughly an hour to get ready for digging after the machine has been got into the field, and another to fit the machine for the road, and this supposing seven skilled workmen are employed; half the time would suffice to put down roundabout tackle in a large field.

"We regret to have occasion to speak thus unfavourably concerning an invention the ingenuity of which we cannot sufficiently admire; but we have to deal with facts as we find them. It is not Mr. Darby's fault that bad weather might render his digger sometimes unworkable when the steam plough could be at work every day or all day long; or that farmers do not like the idea of a heavy machine working backwards and forwards over their fields. He has proved that it is possible to make a digging machine that digs to perfection. He must now prove that this machine can be worked year in and year out in competition with the steam plough, and that his digger will do better and cheaper work than the plough."

During the trials at Carlisle, the comparative performance in terms of volume of earth cultivated per square yard were taken between McLarens broadsider and a digging plough made by Fowlers, hauled by a Burrell engine. The digging plough possessed five breasts, which replaced the ordinary shares and mouldboards, with breasts which when hauled at some speed turned over the soil, breaking it up in much the same way as a digging engine did.

Taking several samples of earth down to the depth of the hard pan in each case, it was found that Darbys engine thoroughly stirred an average of 479 lb. and Fowlers plough 474 lb. per square yard—an almost identical result.

In August 1880 the same digger was on trial in a field at Scholes. It gave a creditable performance viewed by "practical agriculturists", working the strong land at the rate of about an acre per hour. It was stated that Mr. McLaren had arranged for the sole rights to manufacture the digger and intended to modify the steering arrangements from the rear to the front of the machine and increase the number of fork bodies from three to five (or six). First of these modifications was indeed carried out in later engines, but as we shall see the sole right of manufacture was not retained by McLarens.

As with any steam engine used on rail, road or farm, occasional trouble was encountered of one sort or another. Thomas Darby was no exception to this and suffered at the hands of the Law on at least one occasion.

While travelling through Fuller St., Fairstead on 24th August 1888 a Darby digger steamed past some cottages. It was possibly one of McLarens broadside machines as they were concerned in the matter subsequently.

Some half to three quarters of an hour after the passing of the engine it was noticed that smoke was billowing from one of the thatched roofs of a pair of cottages at the base of a chimney stack. Quickly the flames spread and before anything could be done, they were burned to the ground.

Early in 1889 an action was brought at Chelmsford by a Mr. Walter Wright dealer and publican of St. Annes Castle, Great Leighs, heard by Justice Field and a jury.

Darby denied liability, claiming that as it was some time after the Digger had passed and as a fire had been burning in the grate of the cottage, it could not be his responsibility. In any case, he argued, it was impossible for his engine to emit hot cinders.

The jury took the opposite view and found for the claimant. Darby having to pay the £115 claimed by Wright which, with the costs of the case made him a heavy loser.

A Darby Testimonial Fund was opened in an attempt to reimburse him, as it was felt that he had been harshly treated by the Court.

Darby's defence that it was impossible to emit hot cinders was put to the test when McLaren deliberately tried to make a digger spark, even placing it as near as possible to some uninsured outbuildings in an attempt to set them alight, without success.

Diggers were hired out by Darby frequently. In spite of their unpopularity with farm workers, the same old man who recalled the walking digger, remembered seeing a broadsider dig a fourteen acre field at High Easter in two days. "They took some rare fine harvests afterwards too", he said.

Expensive repair costs, difficulty of working a "new fangled" machine, the very appearance of the engine, were all factors dissuading farmers from buying the engine, but they were keen enough to hire it, leaving the responsibilities of maintenance and repair to Darby. The acreage charge was quite cheap, and was probably insufficient from an economical point of view, merely defraying some of the expenses of his costly project. In a way by hiring, he was defeating his own object. While his engines could be hired cheaply—why buy them? Advantages to Darby would have been experience in the field and the testing of his digger's mechanism, some parts of which wore out with alarming rapidity.

On one occasion six diggers were out on weekly hire. By Friday five had broken down. During the evening there was a knock at the door and the crestfallen driver of the sixth engine reported "The main axle's broken, sir". "Well that's a relief," said a bland Darby, "there's nothing more to go wrong now."

Octavius Deacon described the scene when a broadsider suffered a mishap:

"To him, as to Lord Palmerston, 'a difficulty is a thing to be overcome;' and to see this massive man of over six feet stature in a difficulty is a sight for the gods. Never shall I forget his undaunted labours in the field at Derby, when the digger, in the midst of its operations, with a crowd of critical farmers and engineers standing around, suddenly sank into an old pond which had been deceptively filled with light soil. To all who did not know the determined character of the man, the digger seemed to be irretrievably lost and its reputation destroyed for ever; but its inventor rose equal to the occasion. He moved quietly about from one side of the vast machine to the other, placing a screw-jack here and a block or plank there; and, aided loyally by his faithful men, the huge mass was, after a few hours' labour, again steaming over the field, but sorely strained."

The total number of broadsiders built is in some doubt without the benefit of further evidence.

It is known that four walking machines were made—counting the first and second engines as two (as Darby described) and not as one modified engine. A small experimental engine—the first wheeled one—was next, and an advertisement in 1887 proclaimed that seven diggers, which must have been wheeled engines, were working in the Pleshey, Thaxted and Sible Hedingham areas. Thus it can be reckoned that twelve engines were made. McLarens produced four and Savages five diggers, with one being accredited to Agricultural and General Engineering Co.

It is therefore on record that twenty two broadsiders were made. Unfortunately this figure becomes less clear cut when it is taken into account that Darby bought one of the McLaren machines, also that some of the seven engines advertised as working, could also have been of McLaren manufacture. So the number varies between a probable twenty-two maximum to a possible minimum of eighteen or so. Not very many in the light of the great amounts of time and money spent on them. Only four or five of the total number were actually sold to customers, the rest being let out on hire or development machines.

Darby in an interview in 1886 mentioned that "A company made half a dozen machines and then stopped; and I bought the machines of the company". This was too early for the company to be Savages, and was probably McLaren, who never did seem to be as enthusiastic as they might have been, although, to make six machines was a greater number than any, other than Darby. It seems that the same criticism could be applied to McLarens as to all others concerned with broadsiders, namely lack of sales drive. It is doubtful that The Agricultural and General Engineering Co. was being referred to by Darby as they were advertising broadsiders at the same time as the statement was made. There is of course the possibility that more engines were made either by the known manufacturers or by unknown makers, as many as thirty being built by 1898 according to R. H. Clark. The exact number will, in all probability never be established.

An article in *Steam Farming*, contained an eye witness account of Darby's work, when the unnamed writer paid a visit to his farm in 1886.

"Feeling that the subject was one which called for special notice, I have availed myself of Mr. Thomas Darby's courteous and hearty invitation to visit his own farm of Pleshey Lodge, and others in the Chelmsford district, to see there, on the one hand, the patent 'Digger' at its work, and on the other hand to take for myself, in the midst of the harvest, some general observations on the results of that work. Looking back upon the visit, it is hard to say in what manner a day could have been more profitable or more pleasantly spent. Mr. Darby is most excellent company. He is a good though unassuming talker, and is so thoroughly earnest in conviction of the truth of his principle that one is apt to catch his own enthusiasm. And, indeed, it is not difficult to arouse enthusiasm in contemplation of such a vast area of possible usefulness as by the latest evidences appears to lie before his method of applying the principle. Then, his conversation on all matters concerning the farm and agriculture teems with information such as could come from no man who is not, like himself, at once a practical farmer of life-long experience and a careful observer of the results of scientific research, accustomed to examine those results by the light of practical experience, and to apply—as a rule with success—such of them as after that test might stand commended to his use. Adding to all this the hospitality of Pleshey Lodge, and the pleasure of a drive through a pretty and prosperous looking country on a day of magnificent harvest weather, my description of the day's occupation is left scarcely open to challenge.

"From Chelmsford Station to Pleshey Lodge is, by the nearest road, an hour's quite easy driving. The farm consists of about 250 acres of the heavy clay loam land which is the characteristic soil of the district, and, indeed, of the greater part of the county. The land is in itself of not more than fair average quality. Mr. Darby is tenant also of a somewhat larger farm on the other side of the town, but too far to be included in my visit. A circuitous route from Chelmsford Station was purposely chosen. On the way my host was constantly occupied in pointing out those farms which had come to a greater or less extent, and since earlier or more recent dates, under the steam digging system of tillage. Here was a field which had been regularly dug for several years; there, another that saw the digger for the first time a year ago. In either case the close piling of the sheaves affording to the experienced eye prima facie evidence, at least, that the soil was grateful for the change of treatment. At intervals we paused to make an excursion into one field or another for close inspection of the crop.

"One of the most prepossessing features of the new system is the inventor's quite eager readiness to submit its results to the closest examination and the most trying tests. Every question is frankly and fairly answered, and he put forward no claim that can be pronounced unreasonable, or that is without ready substantiation. The five diggers as yet built and put to work by Mr. Darby, I learned, were all out in various directions, doing their utmost; and the demand for their hire has so far grown that there is no doubt twice the number could immediately find employment. Along the line taken in our drive the fields not yet brought under the system are fast becoming the exception, and the dug fields the rule. 'We haven't much steam ploughing in this district now,' said Mr. Darby, replying to my remarks upon competing systems. 'A good deal of steam ploughing used to be done here, but the digger has beaten them almost entirely out of the district.'

"After this manner, by easy stages, we reached the bounds of Pleshy parish shortly before mid-day. Pleshey Lodge was pointed out, crowning a gentle swell of land at a little distance; and, close by, a small slow-moving cloud of steam marked the spot which was to be the central point of interest in my journey. Parting for a few moments from my host, I took a short cut across the fresh dug field to the place where the giant was labouring, doing, as the figures show, the work of 170 men, or of 10 men and 30 horses. The efficacy and thoroughness of his work was at once made manifest to me by the difficulty I experienced in walking over the much-disturbed surface of the land. Instead of being cut into alternate banks and furrows, the soil was evenly and uniformly broken up, and the quantity of air which had been let into it was such that the whole surface where the machine had passed was raised by some three or even as much as four inches; and it was left so open and so full of interstices that the foot sank deeply in at every step.

'If,' said Mr. Darby, 'I had got on to the field a month or so sooner, once digging would have been ample. There is nothing more to be done to the land after this. I shall let it lie a couple of months as the digger leaves it, and then put the seed in.' Taking a shovel, he commenced to clear away about a square foot of the loosened soil, and even after my experience walking across the field, I was hardly prepared to see that twelve inches of depth had to be cleared out before the lowest point of broken earth was reached. Below this there was certainly no 'pan' of hardened soil. Pushing in his fingers, Mr. Darby showed that the digging had reached a depth where the sub-soil was in its naturally porous state. 'This is a four-acre field,' he next explained. 'We came into it about noon-time yesterday, and you see we are digging this one twice over. I mean, of course, that we brought the machine in at that time. There was some time needed to get it into position and so on before we began.' It was now again approaching noon-time, and it was evident that the work would be finished in some two or three hours more. This, of course, was merely a chance record, and not the work of a machine put on its trial and made to do its utmost. From five to ten acres, according to the fineness of soil required, once dug, is, I am assured, the full day's work of the machine.

Fig. 25. Octavius Deacon sketched this broadside digger at work in the Pleshey area
This is a Darby-Eddington engine of later design, about 1886
(E.R.O.)

"In truth, one might well be excused for hesitating to endorse so abrupt a violation of all the canons and precedents of steam cultivation and steam locomotion, as is embodied in the 'broadside' principle, which is the whole foundation of Mr. Darby's invention, and is, in fact, the subject of his patent. To the unaccustomed eye there doubtless does appear a good deal of uncouthness in the implement (Fig. 25). Like the crab amongst animals, its habit of moving sideways is apt to be a source of ridicule, until, like the despised crustacean, it proves that its gait is the one best suited to its particular walk of life. It is a large machine, but by no means unreasonably cumbrous. Beyond all question, in this last respect it is by far superior to the simplest of steam ploughing tackles. It is quite self-contained. When its work is done, there are no wire ropes to wind, no windlasses, anchors, nor tackle to be collected and stowed for transport. By a simple yet very ingenious arrangement of the gearing the digging forks are disconnected, and the machine, consenting to be divorced from its wonted sidelong movement, astonishes the unprepared observer by moving off end-foremost on the self-same wheels, at a travelling speed of three miles an hour, and taking in tow the fore carriage provided to support it when it is moving broadside on. On the farms in this district a large proportion of the fields are neither level nor of good, regular shape. Yet the digger travels into every corner, and seems to work up or down a slope with as much facility as if it were all level land. In these respects, and especially in its habit of finishing up before it leaves the field, the digger scores substantially over the steam plough; for it is a common practice after steam ploughing to send in horses to clear up the difficult corners."

According to this commentator there were five diggers working at this time, mainly on hire to neighbouring farmers.

"No two of them are exactly similar, for in each fresh case it has been found possible to introduce improvements, always tending towards greater simplicity. That which I had now the opportunity to observe was built in 1881. There is no excessive noise, no stoppage, and, so far as I could see, no difficulty whatever. The driver and the guide sat comfortably aloft, and seemed to have a very easy time of it. The former, in fact, was able for several minutes to leave his engine at its work in order to take charge of our horse and dog cart.

"Leaving the great engine at its toil, we passed on through the fields towards the house. 'Look at this wheat,' said my guide; 'I have put on no manure for it whatever. The manure is put into the land for me by the atmosphere.' It was unquestionably a very fine crop, in excellent quantity and condition, and my response was simply to congratulate the farmer on the large supplies of this most excellent manure everywhere ready to his hand. From the wheat I passed into a first-rate field of clover, to which my earlier description of a similar crop identically applies. 'You have seen,' said my host, 'the digging on the clover land and the clover before it is taken off; now you see the result of the wheat crop from the land dug after clover.' Satisfied that I was well impressed with these three stages of his course of farming, he then led the way to where a new digger—the latest type of all—stood waiting the completion of its trials. On this machine there are six forks instead of three, their total width amounting to 22 feet, against the 20 feet of the three-fork machine. Many improvements were pointed out. The machine is considerably lighter. The six forks act more steadily than the three; and the noise and jarring entailed in giving that most useful jerk to the fork prongs are reduced to a minimum by shifting the position of the pin and spring, before described, to a point below the crank and close to the work itself, instead of at the topmost end of a rather long fork 'handle'. This new machine was standing in another wheat field, bearing a crop as like as possible to the one we had just left.'

The business side of Darby's work was somewhat chequered. Engineer and farmer, he perhaps concentrated too much on perfecting his broadsiders, and while surrounded by willing friends he left the selling of his engines to the same friends, relying on the performance of his machines as their best advert. This may explain why the range of working was confined to Essex, more particularly in the Pleshey area and to the district around Kings Lynn, Norfolk, where in both areas, the engines could be seen at work and hired by local farmers. In eight years some 9,000 acres were dug mostly if not all on hire work. Praise regarding the quality of work poured in to Darby. "The work done upon my farm by your digger is without exception the most perfect piece of Steam Cultivation I ever witnessed. The difference in the appearance of the field now and seven days since is almost incredible" said Mr. W. J. Beadel, of Springfield Lyons, Chelmsford in a typical testimonial. Claims of increased yield from land dug by broadsiders were made. Acres dug in this way produced a 20% increase in the yield of wheat.

There is no doubt that the walking diggers could never have become a commercial proposition. Enormous technical trouble with the first four walking machines had led Darby to design the wheeled engines, which were a great improvement, technically and economically capable of cultivating land.

Darby's pioneer walking engine of 1877 was financed from his own pocket, as were the buildings erected for the purpose of manufacturing his engines. The "trifling" amount of £600 for this engine did not include the cost of subsequent modifications. In 1878 the third walking engine proved to be a costly failure. As previously mentioned cast iron components "shivered to pieces" at a cost of £1,600 and the engine was abandoned.

In 1879 Mr. G. C. Phillips, a surveyor from Chelmsford, met Darby and they soon became firm friends. In the same year they formed Darbys Patent Pedestrian Broadside Digger Co. Ltd., with offices at 60 New St., Chelmsford. Darby was appointed Manager and Mr. G. B. Hilliard, Secretary. The Company "composed of some of the best people in Essex", raised £3,500 to further the cause, and the fourth engine—the last walker—was made and shown at the celebrated "Mud Show" at Kilburn, at a cost of £1,900.

This engine too, was abandoned and an unspecified sum of money was voted to develop the idea and a "small trial machine" was made to test the improved form of digger which was now entirely supported on wheels. It was found to confirm expectations of performance.

In a lecture, Mr. Phillips described subsequent events:

"Unfortunately at this point their 'hearts failed them for fear'. The funds were getting low, and no further shares were taken up, so they thought it would be best to wind up by private liquidation, and wash their individual hands from any risk of personal liability. Once again we were thrown on our own resources, and were not long before a traction engine builder at Leeds undertook to build and exhibit a machine at the Royal Agricultural Show at Carlisle."

This was the first McLaren engine of 1880.

"It was not long after this that a London Company was formed to purchase the patents and develop the machine. With the usual flash-in-the-pan and sounding of trumpets it bid fair to rise like a mighty meteor in the agricultural sky, which was then only beginning to get cloudy and tempestuous. But, alas for the frailty of all human arrangements! the usual fate of young public companies, hastily and imperfectly organized, soon followed, and a sorrowful period of darkness and oblivion followed the suddenly extinguished brightness of this star of hope to the benighted agriculturist."

This was the Patent Pedestrian Steam Digging Co. Phillips in his lecture, given in 1886 went on to say:

"Without touching upon any further details, it is sufficient to say that we were again thrust upon our own resources; and nothing daunted by reverses and failures, we set to work quietly and steadily to prove the value of the machine as a practical agricultural implement. Mr. Darby had all along farmed 240 acres of land, little bits of which had been dug from time to time by his experimental diggers. He now determined to go thoroughly into steam digging, and, like a good and conscientious doctor, he took large doses of his own physic before he prescribed it for other people. When the last company's assets were sold by auction, we managed (somehow) to buy the whole of the digger stock-in-trade, and went steadily to work to get up a digging business. At first, Mr. Darby was himself his best customer. Most of his neighbours looked upon it as a very risky venture, and stood carefully aloof. This drove us into another line of tactics. We managed to hire more land, and so extend the digging business, by becoming larger customers ourselves; 300 more acres were added to the previous 240. Year after year passed, and crop after crop was grown; people looked over the hedges and began to think they would give it a trial. Customers began to increase, and they were all well satisfied both with the work and its results, until this year we have had six diggers in constant work for five months, and some people have had three times as much work done as they at first intended. In some cases we have had difficulty to get the digger away from a farm, where it had once made a start; and in one instance the farmer offered to pay 5s. an acre beyond our regular price, if we would but do some more work for him before going to the next customer."

CHAPTER FOUR

Last of the Broadsiders

ANOTHER COMPANY to undertake the manufacture of Darby Diggers was Savages of Kings Lynn, Norfolk.

Again, the double ended boiler was used, but this time with six forks arranged along its length. Very similar in appearance to McLarens, the first digger of Savages, "The Enterprise" appeared in 1888 (Fig. 26). The single cylinder of $9\frac{1}{8}$ in. dia. by 12 in. stroke utilised as many parts as possible from their standard traction engine, including cylinder, motion and governor which was two speed.

As with McLaren's engine "The Enterprise" was classified as an 8 H.P. machine, and was equipped with an outrigger.

The two outer forks each with eight tines and the four inner forks each with seven tines, were operated through a series of bevel and chevron gears, these

Fig. 26. An 8 H.P. Darby–Savage digger hard at work. Note the steering wheels deep in the soil. The pensive looking man walking along with the engine is blissfully unaware of the whizzing flywheel. Diggers would be something of a safety inspectors nightmare today, yet no reports of any working accidents have come to light

last being used by Savages in an attempt to quieten gear noise—which on all broadside diggers must have been considerable in view of the numbers used— and were cast complete with teeth as were all gears on Savage's engines. To avoid the awful task of machining the teeth, the gears were assembled in a rig and run together in a tank containing sand and water gradually being brought together over a period of hours until they ran together smoothly.

Three "Enterprise" type 8 H.P. engines were made at Kings Lynn between 1887 and 1888.

The specification included the following data:—
   Weight of engine complete 15 tons approx.
   Grate area 5·7 sq. ft.
   Fork tines 6 in. apart.
   Average depth dug 9 inches
   Average R.P.M. 150.
   Area dug per hour (advancing at $47\frac{1}{2}$ ft./min.) 1·19 acres.
   Average indicated H.P. 24
   Boiler Working Pressure 120 p.s.i.
   Cultivation width 21 ft.

Some extra items included by the makers were a small winding drum, two speed governor pulley and a tubular bar between the chimneys for tossing the cover over at the end of the day.

The remarkable similarity between the details of McLaren's and Savage's engines must be more than coincidence. It is certain that they were both working to a specification supplied by Darby, based on his experiences with his own and Eddison built machines.

It seems that there may have been some interchange of parts between the two Companies. Both Burrells and McLaren supplied boilers to Savages and it is just possible that McLaren supplied these items for the 8 H.P. diggers. Their experience with their own machines would have provided the expertise in constructing the unusual double ended boilers.

Savage's first broadside engine No. 399 was ordered by J. W. Moss of Kelvedon (mispelt Mors in the works order book) in January 1887. He took delivery on 23rd May 1888 and it is this engine resplendently named on its smokebox "The Enterprise" which appears at work in Figs. 27, and 28. It is thought that it is this engine which worked at Terrington in the Fens for a season about 1890 on a large farm.

"The Enterprise" was reportedly used with great success for some thirty years, mainly on hire work around Mr. Moss's farm at Feering, near Kelvedon. He became an enthusiastic advocate of digging and took every opportunity to promote the idea, becoming Darbys best advertisement. When he retired, his clerk succeeded him on the farm. The one time clerk was just as keen about digging, saying that their broadsider was responsible for great improvement of the land in the Layer Marney and Wigborough district, by breaking the long established plough pan, converting sour soil into a fertile tilth never known before.

Fig. 27. "The Enterprise" owned by John Moss

Fig. 28. Another view of "The Enterprise" at work. The taut chains attached to some object out of the picture to the left, may have been a drag harrow or clod crusher to break up the larger clods previously dug by the engine—which is travelling to the right of the picture. Taken in the early 1900s

In 1898 the new style tender mounted Darby diggers—described in the next chapter—were demonstrated at the Feering farm, with continuing faith in the idea of cultivation by digging in that area.

There is no doubt that one of the most valid claims in favour of digging was the breaking up of the impervious pan made by centuries of plodding horses. Depth dug was a maximum of twelve inches, although Savages specified between $6\frac{1}{4}$ to $9\frac{1}{4}$ inches deep. It was of course, not always desirable to dig to a great depth; for example on soil underlain by clay. In this case it would do positive harm to bring the yellow sticky mass to the top. The many critics who maintained that the heavy machine left the soil in a more compacted state than before forgot that the engine dug *after* passing over it.

Another digger is mentioned in *Fenland Memories* by A. Randall. Owned by either Harrison Bros. or James Walker, it was known to be working at Terrington St. John in 1890 and was impressively named "The Juggernaut".

One of Savage's 8 H.P. broadsiders was shown at the R.A.S.E. show of 1899 where it was awarded the Society's silver medal, possibly the last public exhibition of a broadsider.

Frederick Savage (Fig. 29) was almost illiterate being able to barely sign his name, and therefore few of his own words have survived. His uneasiness with the use of words can be seen in his draft—probably for a catalogue—in which he describes the Darby broadside digger as it was made in his manufactory at Kings Lynn. It is included here in full at the risk of repeating some details, but is of unique interest as an insight into the personality of the great Norfolk engineer, and the methods used in operating a broadside digging engine.

Fig. 29. Memorial to Frederick Savage in King's Lynn as it is today

"I am an Engineer and am the owner of St. Nicholas Iron Works, King's Lynn, Norfolk.

"My principal work is making Ploughing and Traction Engines, Steam Cultivating Implements and Threshing Machines.

"I have known Mr. Darby for many years. I have gone into the question of Digging very carefully and in my opinion digging is far better than ploughing either by horse or steam, for this reason, it breaks up the subsoils and the air and water can get down. In this way it answers for Drainage and also the air gets into the land for manure. I took the matter up some few years ago and have made two Diggers.

"I consider the invention to be one of great novelty and merit. I have studied Mr. Darby's Specification and drew my drawings from it from which drawing I manufactured the last Digger.

"I produce my drawings. The following is a description of the digger as manufactured by me. The boiler is of the Locomotive Type but with the Fire Box in the middle of its length. When digging the machine moves side ways, when travelling end ways. The Engine for Driving the machine is placed horizontal on the top of the boiler and drives a horizontal shaft parallel with itself which again through suitable gearing drives a counter-shaft which in its turn through similar gearing drives a long shaft which again drives three shafts which actuates the cranks which give motion to the diggers. At the same time two vertical shafts driven from the primary longitudinal shaft give motion to the four bearing wheels and so enable the machine to advance. The Diggers are provided with steel tyres which are fixed by bolts and locking plates to a cast steel or cast iron arm which is hinged into a cast steel or cast iron lever which is supported at or about the middle of its length by a radius bar and is actuated at its upper end by a crank or eccentric so that the lever and the tyre have a vertical motion and run round the radius bar very much the same as that which is given to a spade in digging. The cranks are so arranged on the several shafts that the diggers follow one another in their action. When digging the machine is provided with a separate carriage which is attached to one side of it and is used for steering purposes and which also serves to regulate the depth of the spit cut by the diggers the centre line of the boiler being regulated vertically with reference to the carriage by means of a hinge and capstan bolts. When it is desired to travel along a road or from one field to another the bolts which hold the bearing wheels in place are moved and the same power is used for advancing the machine serves to turn the wheels so that their axis is across instead of under the boiler the tyres are removed from the centre holders and are only left on those at the ends and in these they are turned on the bearing which supports them until they are in a nearly horizontal position the wheels are again locked into place by the bolts the steering carriage is removed from the side and is attached to one end of the machine and serves as a tool box and the Engine is ready for travelling along the road and from place to place. The operation of altering the machine from its digging form to that of travelling occupies two men about 20 minutes. This operation consists in releasing 8 springs and removing 6 bolts which hold 6 of the 8 forks in position the driving shaft of each end pair of the forks is then lifted slightly by means of a union screw and two bolts are removed from the framing of each so as to allow the 4 end forks or diggers to be thrown out of gear. 4 bolts which hold the carrying and propelling wheels or rollers in position are removed one of the end bearing or travelling wheels is disconnected from the shaft by means of a spring bolt the power of the engine is coupled by the ordinary clutch into one of the vertical shafts which communicate motion to the propelling wheels and one only of these wheels being in action a turning motion results the axis of the wheels is gradually brought from a position which is parallel to the axis of the boiler to one which is across it and the bolts which hold the frame in position under the smoke box are replaced. The next proceeding is to disconnect the steering truck from the boiler by the removal of 2 collars and 2 blocks. When this is completed the same procedure is gone through with the front pair of bearing and propelling wheels as was done with the hind ones with the exception that the bolts which hold the frame in position are not replaced but a pair of chains is attached to a semicircular frame and these wheels are then used for steering purposes when these operations are completed the engine is on a position to travel endways instead of sideways as is the case when it is digging. When in this position the machine can be used for thrashing purposes chaff cutting and for other farming purposes for which the steam power is necessary.

"Having a great deal to do with steam ploughs and also having studied cultivation by Digging generally and especially by Mr. Darby's Digger I am in a position to judge as to their respective merits.

"The cost of digging an acre of land with the digger is considerably cheaper then that with the plough especially considering it with respect to the results obtained.

"An owner of a digger can dig an acre of land with Mr. Darby's Digger for 6/– per acre and I consider this to be by far the cheapest mode of cultivation now open to the Agriculturalists.

"The cost of digging an acre as before stated is 6/– as against 8/– for ploughing.

"I have fully acquainted myself with Mr. Darby's specification of patents and am of opinion that the invention is properly described in the specification and that the main invention is described in the specification which is now sought to be extended and that the other 2 patents are for Improvements in detail the one of 1881 being found not to be useful was allowed to drop.

"I consider the machine to be one of the greatest practical value to the Agriculturalists and if now pressed forward and backed up by influential men with Capital will become a great boon. I attribute great merit to the broadside principle because such a large space of ground can be dug up at one time.

"The following are a few of the advantages gained by Mr. Darby's Digger over the other systems of Steam Cultivation.

"Signed
"Frederick Savage."

Savage has described an eight fork engine, while drawings of the 8 H.P. diggers and photographs of "The Enterprise" all show a six forked machine. However, the principle of his description is unaffected.

How a broadsider in road travel setting was steered is something of a mystery. Presumably the leading pair of wheels steered in exactly the same way as a traction engine, but the writer has not found any photographs of a digger being driven on the road to give any clues. Drawings, including Savages works drawings do not include a steering column and worm to the engine wheels, only to the outrigger. One becomes even more puzzled when it is taken into account that it depends on which way the engine travels lengthways as to which wheels steer. To which end is the towed outrigger attached, and where does the driver stand? Assuming that some form of steering must have been provided (Savage's letter gives no real clue) the mind boggles at the prospect of driving the digger with outrigger trailing, out of the field, through the gate and on to the road! Over thirty feet long, with fork mechanism and other pieces of metal projecting out from the side, it is no wonder that it was an unpopular engine to move.

The work involved in changing the engine from broadside to lengthways modes was considerable. Although Frederick Savage said that it took twenty minutes, it is difficult to see how it could be done this quickly except perhaps under ideal circumstances which are rare on the farm. Moving the unwieldy outrigger and wheels—after unbolting hem—which weighed many hundredweights, must have taken not only a considerable time, but the labour of many men, belying the one or two required to operate the broadsider when digging. Add to this cloying mud on a winters day with the wind howling and rain lashing down, and you have an almost Herculean task for those poor men to perform and a good reason for them to regard the engine with some hatred.

First hand experiences with Darby diggers are remembered by only a few today.

Mr. Arthur Johnson, retired Works Manager of Savages, to whom the writer is indebted for much of the information on their broadsiders, recalls as a lad seeing an 8 H.P. machine proceeding through the South Gate of King's Lynn one evening in 1898. He thinks that this engine was the one owned by James Walker.

Oil lights were attached each end of the boiler for road use and possibly nightwork.

Mr. Johnson says:—"The two head lamps right and left were made from sheet copper, would I think be common to the Diggers, the two I saw in the stores were probably off the 6 or 8 H.P. Diggers, they were of massive construction, rectangular glasses, bevel edged, with guard bars on front, oil wells about the size of the present road lamps, a tiered ventilator was under the steel lifting handle." These lamps survive somewhere as Mr. Johnson remembers them being sold to an antique dealer by Savages. One wonders where they are now, as they are probably the only surviving parts of a digger.

It must have been an impressive sight as the engine rolled through the narrow central arch. Frederick Savage's statue, one of the few raised during the lifetime of the subject, stood on the right hand side of the road, just through the arch. No doubt the driver glanced to one side in acknowledgement, as his engine rumbled through the narrow streets on its way home. This engine too, had been working in the Fens and may have been on its way back to Savages works for overhaul or repair. Mr. Johnsons clearest recollection of the rare sight is the glow from the twin chimneys in the dusk as the regulator was opened and the single cylinder with its twin exhausts, barked and echoed through the archway.

The Reverend R. C. Stebbing also recalls a Darby broadside digger actually at work. The engine was working in Essex and was our old friend "The Enterprise". He writes:

"My reminiscences of the Moss digger simply concern the strange appearance of the chimneys each end. I have a picture in my mind of this unusual sight as the machine came down the field (for years I thought I had imagined it). As I was not more than four years old at the time I took no note of technical details! My father did tell me the digger discovered a spring in the field in which it was working and some delay was caused until it could be extricated. The existence of the spring was known, I think, but not its exact location. The incident showed up a danger to which diggers were liable, horse drawn implements would not necessarily have sunk in."

Unlike other broadside digger makers, Savages made a further attempt to develop the idea and considered it a worthwhile project to make a smaller, compact machine which was very easy to handle, manoeuvrable, cheap to make and capable of travelling about without the awkward job of changing wheels around and disconnecting the outrigger.

Such a machine was designed and classified as a 6 H.P. machine, making it the smallest of any of the broadside diggers made (Fig. 30). Strangely, there is only one surviving photograph, and no records of its success or failure, no clear idea what happened to the machine or to whom it was sold. No trials, shows, comments complimentary or derogatory tell us of its fate.

In one respect however, the engine is better known to us than any other broadside digger, through the existence of the almost complete set of working drawings from which the engine was originally made.

As with "The Enterprise", as many parts as possible were used from standard production traction engines. Of typical broadside digger appearance, this diminitive version possessed a boiler only 9 ft. long overall, including smokeboxes, of standard double ended digger pattern.

The single cylinder, 8 in. by 10 in., stroke and motion were contained within the length of the boiler, and in between the width of two plate steel horn-brackets.

To enable the engine to be driven broadside on through a field gate, great care was taken to minimise the length of the machine. Demonstrating this point is the section of the two chimneys, which are oval. The flywheel projects 6 in. over one end of the boiler, but would have been above normal gate

Fig. 30. A unique photograph of the Darby–Savage 6 H.P. broadsider at work, probably in fenland. This was the smallest Darby digger constructed
(*Model Engineer*)

height, ensuring free passage.

Digging width of the engine with its five sets of forks was 13 ft. 3 in. To reduce the width to a minimum for road use, the two outside sets of forks complete with bearing brackets could be removed. The job was made easier by the davit with block and tackle which can be seen in Fig. 30 which enabled the heavy parts to be lowered to the ground after they were unbolted from the engine. When one fork had been removed, the davit was lifted out of its loose socket, replaced at the other end and the second fork removed. The five throw digging crankshaft was split at each end and it was necessary to separate bolted flanges at each end to release the outer cranks. Effective width of the engine was thus reduced to the length of the boiler which was 9 ft. and could easily be driven through a farm gate. However, the time taken to dismantle and reassemble the outer forks must have been considerable—the writers estimate is 45 minutes—thus detracting somewhat from the setting up time advantage gained by not having to swivel the wheels through 90° as on the larger 8 H.P. engines.

Due to the cylinder being at one end of the boiler, it was considerably nearer one chimney than the other. The exhaust steam being taken from each side of the piston in usual manner, there was a time lag at the chimney between the exhausts due to their unequal length. To even the beat out a butterfly valve was placed in the short exhaust, adjustable by means of a hand turned key to give equal pressure in both exhausts.

The chimneys were made in the round, and then formed into their oval shape on mandrels, the circumference having been carefully calculated before rolling.

The usual outrigger—or in Savages terminology "land cart"—was incorporated. Detachable as with its larger brothers, the land cart was of the similar design and size for both 6 and 8 H.P. diggers. It was adjustable by means of a screw jack about a pivot in the lower part of the outrigger, which tilted the whole engine one way or the other, the object being to raise or lower the forks in and out of work. Steering was by means of four flanged wheels, mounted on a stub-axle, turned by hydraulic ram or chain, both versions are shown on the drawings.

Unlike "The Enterprise" with six forks, the little digger had five. Seven tines on the middle fork, six and five tines on the outer and inner forks respectively. Three variations of forks and operating mechanisms are shown on the drawings. Taking the dates of the drawings as the order in which they were designed, the earliest type is shown on the works specification drawing of February 1889 and consists of acutely angled fork tines, operated by a crank which hinged around a tubular linking attachment. The drawing shows the hinging point within the radius of the road wheels—an impossible arrangement which we can safely assume was not used. These forks were preceded in work by cultivating tines, which would have helped to break up the soil for the forks.

The second design of fork dated April 1889 is shown with half elliptic spring loaded tines—the cultivating tines now being omitted.

Both of these designs suggest that some difficulty was experienced in establishing the best layout, perhaps due to the small engine not being heavy enough to penetrate the soil under certain conditions, possibly even stalling the engine. This view is borne out by the inclusion of a friction device incorporated in the crankshaft which would have slipped if conditions became too severe, or perhaps stones were encountered.

Thirdly and finally, in 1891 a set of forks shown on an inked and colour washed drawing—indicating finality of design—are of quite simple construction, without springing or extra tines.

Plotting the movement of the tips of the forks provides an interesting comparison of the various ideas. The motion of the third design is about as near to hand fork movements as it is mechanically possible to achieve, and is borne out by the photograph as the type of fork actually used.

Two 6 H.P. diggers were made. An entry in Savage's order book as No. 469 being described as a "traction engine for Darby's patent digger", works job No. 437 was the first machine. According to their system if it was entered, it was made, and as no customer is shown, this was presumably a prototype for development work. Arthur Johnson remembers that it was in fact sold later, but as it was before he joined the Company, he does not know to whom.

He also recalls the second 6 H.P. broadsider in Savage's works, which for an unknown reason, does not appear in their records. He did not actually see it in action, but remembers that it was used for demonstrating its capabilities. It seems that it was to no avail, however, as it was broken up in the early 1900s.

The engine was not entirely wasted. In the good tradition of making use of every part possible, the common parts with the six horse power traction engine were used as spares on other engines and the boiler was used for heating the workshops, fulfilling this humble role for some years.

Mr. Johnson is certain that no further broadsiders were made and that the second one was—as with the first—a single cylinder machine, the only annular compound digger produced by Savages being the Darby-Steevenson digger described later.

The wheels of the little digger could not be turned through 90°, indeed, due to its small width it was unnecessary. This engine was the only broadside digger which could be used in the field by simply driving it into position and within a short time commence work—a big advantage over its more unwieldy brothers. Only 14 ft. in overall length and about half the weight of other broadside diggers, thus overcoming many of the criticisms, it is a mystery why the engine was not a success.

Savage's terminology of the otherwise known outrigger as the "land cart" suggests that unlike the larger engines towable extension, it may have been intended for use only on the land. Two columns or legs on the land cart could have been used to support it when detached from the engine, a process made easy by the design. The advantage of this arrangement would be that the engine would be only 9 ft. long and 9 ft. wide after removal of the outside forks. It would have been more easily driven and stored when not in use.

The removal of the land cart presents only one problem for the 6 H.P. engine. Quite simply, due to the unswivelling wheels the engine would fall over! Therefore the implication is that some extra wheels were fitted before removal of the land cart to enable the engine to proceed. No drawings have been found to confirm this theory, but it would seem to be the only one that fits the general design features.

A development of the land cart in the form of hydraulically powered steering was drawn, but probably not included in the engines constructed. The photograph shows a manually operated shaft and worm drive steering mechanism. It would appear that one size of land cart was used for both the 8 and 6 H.P. diggers, but the steering arrangement on the 8 H.P. engine was of chain type as used typically on traction engines and as used by McLaren and Darby, the chains being attached to a yoke which turned the wheels one way or the other.

The large numbers of gears used in transmitting power and motion from the engine to the forks and wheels is probably unique in the design of self propelled steam vehicles. There are no less than thirty six gears on the 6 H.P. digger. Trains of spur, bevel and chevron gears seemed to be the only way the designer could keep the engine compact in size, but the loss of power through the many right angles must have been considerable. When the engine was digging, that is, when the gears were arranged to drive both the forks and inch the engine along as well, no fewer than fourteen gears were turning in one train between the crankshaft and wheels.

Fig. 31. One of Savage's machineshops. Although taken in the 1920s the machines had altered little from Victorian days, and some of the men no doubt worked on broadside diggers

Parts of an annular compound engine, drawn in 1898 for the six horse power digger indicate that work was in progress at this late date. It was not constructed but must have been about the last work to be done on any broadside digger. By this time Darby's Quick Speed Digger and Cooper's machine—mentioned later—were being developed and produced.

Whether Darby himself was personally involved with Savages will probably never be established but as we know from his letter, Frederick Savage and

Fig. 32. "Near side" view of the 2″ scale digger

Thomas Darby were friends of long standing. Savages certainly tried hard to produce a workable machine (Fig. 31).

As an intriguing venture into both historical reconstruction and model engineering, the author has built—from the works drawings—a $\frac{1}{6}$th scale model of the Savage Darby 6 H.P. broadside digger. This is the only broadsider now known to exist—albeit a small one (Fig. 32).

Briefly, the construction was started in late 1975, and was completed in about twelve months. However, before manufacture of parts could be started it was necessary to produce a set of scale drawings. As previously outlined, some of the parts as drawn by Savages are in several versions and decisions had to be made as to the actual version used on the digger. For this purpose, the photograph of the engine from the *Model Engineer* proved invaluable, although of course there were some problems left as one view is unable to show the opposite side, making some parts frustratingly just out of view! Worse, a close inspection under a magnifying glass revealed only hazy outlines of one or two parts in question. Eventually, the details were established and the scale drawings were far enough advanced for construction of the model to start.

Initially, it was necessary to make the patterns from which the basic castings could be made. These were completed in batches and sent to the foundry. When completed the total number of castings to the set was 92. Machining of the castings then proceeded.

The first assembly was the land cart, followed by the main platework around the boiler, to which was assembled the five forks and digging mechanism. The copper boiler, mounted between the plates with the cylinder, motion and two chimneys on top, together with the wheels and drive mechanisms below, completed the general construction.

The model as shown here is unpainted and lacking the davit (Figs. 33 and 34). For comparison refer to Fig. 30. Being unpainted the model is more "photogenic" as a glance at any of Fowlers works photographs of their engines in primer finish will show. Painting tends to hide details otherwise visible, particularly in black and white photographs.

The little digger has been steamed and valuable experience gained in using it.

A confliction of information on the original Savage drawings was discovered when it was found that one of the two main wheels is permanently in gear through to the crankshaft. The other wheel possesses a dog clutch—which frees the wheel from the "dead" axle, enabling the engine to swing round at the end of a digging bout. Savages original drawings show the dog clutch on a half section elevation of the engine, omitting to say if a second clutch was incorporated on the other side. If two clutches *were* built in there would have been a great danger of the full size engine running away if they were both out at the same time. For this reason, it was decided to incorporate one clutch only on the model. However, the other fixed wheel is a distinct disadvantage when moving the model without steam pressure, as it has to be lifted (180 lb.) or put on to a special transporter trolley.

Attempts to dig with the model have proved that there is ample power

Fig. 33. A view from above showing the vertical shaft to the wheel drive gears and the land cart wheels in steering position.

through the train of gears from the 1½ in. diameter cylinder (Fig. 35). In the hot dry Summer of 1976, the soil was so dry and dusty that when digging, the wheels spun and dug in until the land cart supporting legs also penetrated the ground, holding the engine fast with forks flailing! A further trial on harder ground proved more successful, with the forks penetrating to the full depth of 1½ in., tumbling the soil quite efficiently. While digging, the bark from the engines twin chimneys showed how hard it was working. It is quite hypnotic to watch the action of the forks which dig in the sequence 1-4-2-5-3 when facing them.

Lubrication of the bearings must have been almost a full time occupation on the full size engine, with some 45 oiling and greasing points.

A feature unique to the 6 H.P. broadsider was the incorporation of an overload device in the crankshaft (Fig. 36). Effectively, it is a slipping clutch which operated if the forks met an obstruction. This no doubt saved some bent crankshafts and probably took the place of the sprung forks mentioned previously and may have been the result of experience with the 8 H.P. engines which could have suffered gear or crankshaft damage while digging.

Some 1,600 hours were spent building the model and it has proved a re-

Fig. 34. The model digger can be compared to the similar view of the full size engine in Fig. 30

Fig. 35. The digger at work during one of its early "trials"

Fig. 36. Driver's eye view of the Darby digger controls

warding exercise, enabling some of the difficulties experienced by Victorian engineers to be evaluated (Fig. 37).

Broadside diggers may be as extinct as dinosaurs, but just as there is the possibility that the Plesiosaurus of Loch Ness will be rediscovered, so there is the chance that a digging engine may once again be re-born.

With the existence of the original drawings and—as related later—the patterns, together with the knowledge regarding construction obtained from the model, the author hopes that one day a way will be found to construct a full size replica engine.

Advantage of the almost complete survival of the historical material of Savages has already been taken by the construction of two chain drive traction engines of 1870 by Belmec International of St. Germans, near King's Lynn under the supervision of David Bretten.

The arguments for and against re-construction of full size engines are outside the scope of this volume, but in this writer's opinion, where there are no existing examples surviving there is a strong recommendation for a replica to enable the engine to be seen in three dimensions, to evaluate is performance and efficiency (or lack of it). Its existence would provide a historical and educational facility not possible without it, and lastly, speaking as a steam enthusiast, the sheer enjoyment of making it would be reason enough!

Due to the unfortunate disappearance of many types of machinery during the last world war, a comparatively small number of historical engines survive. These are increasingly sought after with consequent escalation of prices out of all proportion to their real value both historically and intrinsically.

Fig. 37. A 2" scale driver demonstrates the close proximity of the flywheel and some gears to the manstand, placing him in constant peril while firing and driving the digger

There is a strong case for reconstruction provided certain qualifications are fulfilled, among these being as strict adherance to the original as possible and enough information gathered to ensure that the resultant work is accurate. The absence of a prototype provides the best reason of all for reconstruction.

According to these conditions, the 6 H.P. Savage Darby broadside digger would seem to be an ideal candidate for an historical exercise of great interest.

It would be fascinating to once again see this unique vehicle working in the field, the clods of earth tumbling over, smoke billowing from the double chimneys, the engine, with its cylinder and motion working hard amid ringing and whining gears. The chance is there to bring alive one of the most complex mechanisms ever devised for steam cultivation.

The sad closure of Savages of King's Lynn on 31 March 1973 was brightened somewhat by the last minute intervention of Mr. W. McAlpine in saving what is probably the most complete set of records and foundry patterns of any concern of this type. This admirable contribution to the annals of steam engine history was further enhanced when Mr. McAlpine presented all the foundry patterns for permanent preservation by King's Lynn Museum. They are stored in one of the huge wharfside Council owned warehouses of Kings Lynn docks, which unfortunately is not open for general viewing. The writer, however was privileged to have an opportunity to see the interior of the warehouse in the company of Mr. Johnson in an attempt to locate the patterns from which the 8 and 6 H.P. diggers were made.

Fig. 38. Some of the patterns from Savage's foundry, now stored in a King's Lynn dockside warehouse
(*Eastern Daily Press*)

The patterns (Fig. 38) are stored in the gloomy stone building in rows on the floor and on benches. The single small window in the roof allowed a few small rays to penetrate one corner of the huge echoing building. It is so dark that a torch was required to see under the benches. Patterns appeared to be in piles everywhere—thousands of them from tiny brackets to huge cylinder blocks, wheels and gears. On investigation, the patterns were in fact arranged in sections each containing the patterns of a particular type of engine and each classified with a prefix letter to the pattern number, signifying the particular type of engine to which the pattern belonged.

Mr. Johnson found some patterns made by himself. His father was foreman pattern maker to Frederick Savage.

The search for the digger patterns was rewarded after some while by the writer recognizing one of the large bevel gears from the 8 H.P. machine. Next to these were some 6 H.P. patterns and on checking through them it was found that some of the set were there, and a thorough search will hopefully find them all. The patterns are made of Siberian Yellow Pine and are generally in very good condition.

In closing the story of Thomas Darbys broadside digger's, it may here be appropriate to return to the early days of his work.

With the formation of the Darby Land Digger Syndicate Ltd., all attempts in the ensuing years produced very little commercial activity. In spite of the generous publicity and the vast amount of technical development, sales were virtually non existent. Even the efforts of well established concerns like

Savages and McLarens produced few orders. Diggers were too speculative, not to mention expensive. Conservative farmers favoured traditional methods of cultivation, and were even prepared to use the time honoured plough—albeit modified—in conjunction with steam ploughing engines, but baulked at the idea of this revolutionary method of cultivation, resulting in poor sales for the Syndicate.

Between 1877 and 1898 the enormous sum of £100,000 was spent. A ten year extension to the patent granted in 1891 did not help the situation.

This was not the last to be heard of Darby, however, and with the commercial failure of his broadsiders, he introduced a new and completely different type of digger, bearing no relation whatever to his previous machines, but was of the same basic type as several other inventors ideas.

CHAPTER FIVE

Darby Tender Diggers

DARBYS NEW DIGGER was known as the "Quick Speed Digger", and took the form of a vee shaped attachment to the tender of a standard type traction engine (Fig. 39). Compared with the broadside diggers this was a simple and cheap mechanism.

The principle was now rotary, and consisted of numbers of rotating arms with enlarged ends arranged vertically along the two arms of the vee, and driven by shafting from the engine itself.

Attached behind a reinforced tender, the first machines dug a swathe 14 ft. wide, which by 1898 had been reduced to a width of 11 ft. 6 in.

In this year *The Engineer* reported "Mr. Darby is indefatigable in his endeavours to produce a satisfactory digging machine, once more he is literally in the field. . . . The outcome of many years of study and experiment to produce a device which shall be capable of digging the land more cheaply and efficiently than has hitherto been done by a plough and horses. The digger weighs $2\frac{1}{4}$ tons and is supported on its own wheel so as to raise and fall independently of the engine to ride over unequal ground.

In 1894 an engine with digger attached was sent for exhibition in Hungary. It must have made some impression as six years later in the *Implement and*

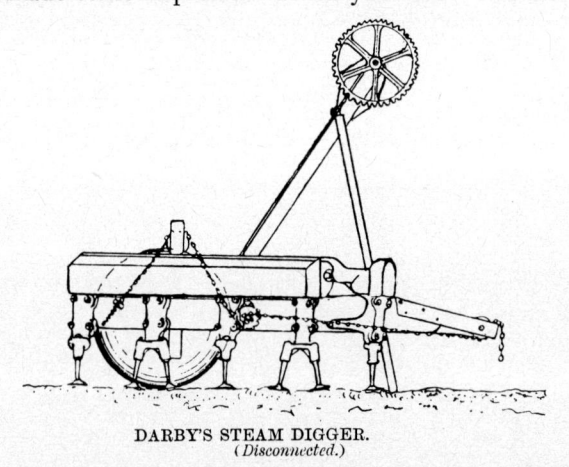

DARBY'S STEAM DIGGER.
(Disconnected.)

Fig. 39. With vertical rotary twin arms this new type of Darby digger could be detached from the engine which could then be used for other purposes (*M.E.R.L.*)

Machinery Review it was reported that "The Darby Land Digger appears to have a radiant future in Austria and Hungary. In a letter from Buda-Pesth Messrs. Nicholson there make it known that Count Andor Zichy (one of the best and most practical farmers in Hungary) considers its work 'admirable'. Its simplicity is wonderful he adds and if it can be made to dig 12 in. and the subsoil brought to the top there 'is a great future for it in Hungary'."

One somehow has the feeling that Count Zichy was not perhaps as enamoured with the digger as Messrs. Nicholson would have us believe and with some imagination perhaps we can see the Count giving an unenthusiastic twirl of his moustache and mumbling into his beard "admirable" whereupon Messrs. Nicholson seized upon the magic word as if praise had been heaped upon Darby's latest wonder. The digger was obviously incapable of digging to a depth of 12 in., neither did it bring subsoil to the top: both essential as far as some farmers—in Hungary or anywhere else—were concerned.

A company was formed to promote Darbys most recent digger. Named The Darby Land Digger Syndicate, of 6 Billiter St., London, George C. Phillips, maintained his loyalty to the idea of steam digging by being Secretary to the Syndicate.

An advertisement in *The Implement and Machinery Review* proclaimed the tender mounted digger as "The agricultural steam cultivator of the 20th century, destined to dominate the World's food supply", a prediction unfulfilled.

Costing £350, the diggers were used in England and several other countries. They were attached to traction engines made by Ramsomes, Sims & Jeffries, Fowler, Ruston-Proctor, Clayton and Shuttleworth, Wallis and Steevens, and Burrells.

It is not clear if all the Quick Speed Diggers were manufactured by Darby, although wording of advertisements suggest that they were, the various engine makers hoping to sell more engines by offering them complete with digger supplied by Darby. Photographed diggers have the Syndicate's name painted on, with the legend "T. A. & S. C. Darby's Patent. Manufactured at Pleshey, Chelmsford". Thomas C. Darby, Senior had passed his work on to his sons, probably keeping a fatherly eye on the Syndicate, enabling him to develop his ideas and continue farming.

An early version—probably the prototype production tender digger—is shown in Fig. 40. The cast iron nameplate incorporates "No. 1", presenting an appearance of a rather rough and ready machine, dented and unpainted contrasting sharply with later models. Thomas Darby stands by the engine. This digger was made at Pleshey, as was the second, improved digger. They were made in Darby's collection of sheds which were quite unsuitable for producing diggers in quantity.

In 1900 manufacturing was moved to larger premises, Stileman's Works at Wickford, near the railway station, where the business remained until its closure. It was latterly run as Sidney Darby (Wickford) Ltd. by the son of Thomas, previously mentioned.

Fig. 40. Darby's first production tender digger made about 1898. Attached to a Ransomes, Sims and Jeffries single cylinder traction engine

The "No. 1" tender digger possessed semi-elliptical rotating tines to perform the work, gear driven by a train of exposed and potentially hazardous bevel gears. Adjustment of digging depth was by means of a handwheel projecting from the rear of the digger.

Shortly afterwards, the second machine appeared very much improved in appearance (Figs. 41 & 42), complete with guarded gears and shining, lined paintwork. It is this digger which appears in advertisements of both the Syndicate and Ransomes, mounted to the tender of a handsome single cylinder 8 H.P. single cylinder Ransome, Sims & Jeffries traction engine, the complete ensemble having a most impressive, almost elegant appearance.

Undulating ground could seriously affect the depth dug from none at all, to plunging the digging forks to a damaging depth. Cooper overcame this problem by using hydraulic means to adjust the digger height, but Darby utilised a mechanical means of adjustment.

He attached his digger to the rear axle of the towing traction engine by means of a yoke which embraced the tender, reaching round to the ends of the engines rear axle, to which it was attached. A screw adjustment was provided, also at the axle ends to alter the depth of dig. The whole weight of the digger rested upon a single wheel positioned in the centre of the triangular planform of the digger. As it progressed, working variations in the surface of the ground were followed by the wheel thus pivoting the digger about the engine axle centre. This of course, necessitated some form of flexibility in the method of transmitting power from the crankshaft of the engine to the digger (Fig. 43), achieved by shaft drive on earlier models and chain drive on the diggers made

Fig. 41. The second tender digger of c. 1899, in smart finish to match the new Ransomes engine. This was probably the last digger to be made at Pleshey

at Stileman's Works in Wickford from 1900. Darby was among the first to overcome the difficulties of flexible power drive.

Praise was heaped upon the digger's performance. "It is without doubt the most effective steam digging implement that I have seen for use in this country. The work done is excellent", wrote Professor Robert Wallace of

Fig. 42. Rear view of the second machine. The fine tilth produced by the rotary tines can be clearly seen

86   DIGGING BY STEAM

Fig. 43. Adaptability to uneven ground was a useful feature of the Darby tender digger
(*M.E.R.L.*)

Edinburgh University. "I have had good crops after using your Digger, and get better results than after the plough" was a typical comment from a farmer.

Ransomes general catalogue of 1906 included Darbys tender mounted digger. The name of our acquaintance from broadside digger days, John Moss of Feering, Kelvedon, Essex appears among the testimonials. "We have used one of your patent diggers for several years with the utmost success improving every field we have dug with it. We always noticed a marked advantage in all

The Darby Digger

Fig. 44. A digger made at Darby's new works at Wickford after the turn of the century

crops where the land is dug instead of being ploughed." Moss became a devotee of steam digging when he purchased the broadsider from Savages in 1888, and was reputed to have exclusively used this method of cultivation for thirty years. He is a typical example of converts to the idea of digging. It is one of the anomalies of the subject that all who used diggers were enthusiastic in their support for the idea, claiming for their own efforts all that Darby said they would achieve. Yet very few could be persuaded to purchase diggers, especially in the early days. The syndicate fared somewhat better commercially than the broadsiders, selling some thirty Quick Speed Diggers, manufactured at Stileman's Works (Fig. 44), some being exported to Canada, Egypt, South Africa, the Isle of Man, and at least one to Hungary.

Darby and Steevenson (of Eddington and Steevenson) commenced partnership in 1890, when production commenced of two types of digging engine to their specification and by 1893 several had been made and sold.

The "Nottingham" machine was the heavy version (Fig. 45), capable of digging 40 to 45 times per minute to a width of 21 ft.

The "Colchester" machine was lighter and it was reported to have been successful in trials at Newcastle (Figs. 46 & 47). It was so named as it—and the "Nottingham"—were constructed at Colchester by Davey Paxman & Co.

Each engine weighed about 10 tons and could enter an 8 ft. wide gateway.

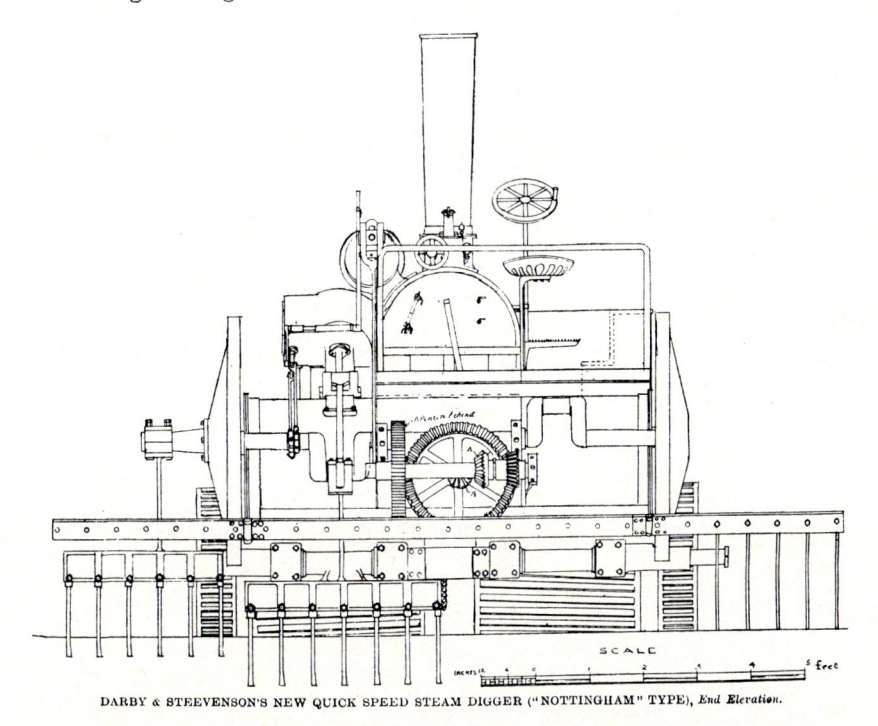

DARBY & STEEVENSON'S NEW QUICK SPEED STEAM DIGGER ("NOTTINGHAM" TYPE), End Elevation.

Fig. 45. The Darby–Steevenson "Nottingham" Quick Speed digger
(Implement and Machinery Review 1895)
(M.E.R.L.)

Two were in operation at Great Caulfield, one of which, ordered by a Mrs. Bailey was named "The Caulfield Enterprise". Other purchasers were the Duke of Buccleuch, Mr. Mason of Swaffham and a "large landed proprietor in Hungary"—perhaps Count Zichy?

The general appearance of the "Nottingham" digger was in some respects not unlike the earlier broadsiders, except that the digging mechanism was now arranged across the end of the boiler instead of along its length. Four sets of forks considerably wider than the engine and having six tines to each of the outer and seven to the inner forks, were driven by a combination of a two throw crankshaft on the inner forks and crankpins on two flywheels mounted on each end of the engine crankshaft, providing drive to the two outer forks. The digging crankshaft was mounted very low down on the tender and driven from an angled cylinder and connecting rod.

Probably one of the most handsome digging engines made, well designed and finished, the "Colchester" possessed two narrow track, angled front wheels, making the ground contact planform a three wheeled engine, able to turn nimbly at the headlands. With a single cylinder 9 × 12 in., like its larger brother, a flywheel was mounted on each end of the crankshaft, which was this time mounted over the firebox. Two massive curved arms, constructed from plate steel were driven from crankpins on the flywheels directly to the two sets of tines, twenty one on each. A second set of what appears to be forks across the rear of the engine were in fact merely fixed tines which helped to break up the dug soil—effectively a trailing cultivator which could be raised or lowered by a hand lever to enable fouled tines to be cleared. In the photograph of the

Fig. 46. "Colchester" Quick Speed digger manufactured by Davey Paxman to the Darby–Steevenson specification

Fig. 47. Offside view of the "Colchester" engine. Narrow front wheels to facilitate turning at the headlands can be seen, together with the twin flywheels and massive crank arms that drive the forks

working digger in Fig. 48 a distinct bend can be seen in the supporting channel suggesting that the strain was excessive at times.

A feature of both the Nottingham and Colchester diggers was that the digging mechanism was built on to the engine as an integral part and—while removable for repair and maintenance—was intended to remain on the engine permanently. Thus the engine was constructed primarily as a digging machine, all other applications being ancillary. A small winch, mounted in the tender of the Colchester digger would be a very useful item to enable it to haul itself out in the event of the engine getting into difficulties while working (Fig. 48).

The two types of Paxman built diggers were made between 1888 and 1894. Darby's Quick Speed tender mounted digger proving to be the better commercial proposition of the two, continuing in production until 1913.

Savages of King's Lynn made an attempt at the Darby Steevenson digger by constructing one engine of this type. Works number 507, it was an annular compound engine $8\frac{1}{8} \times 12\frac{1}{4}$ in. with a 14 in. stroke making it the largest cylinder made by Savages for any of their many engines. This digger was made in 1890 and its fate is unknown.

Somewhat indefinite details of an engine known as the Darby Maskell digger were mentioned in a letter from McEwan Pratt of Wickford, Essex to Baguley Cars Ltd. Dated July 1912 the letter refers to the transfer of work to Baguley Cars and to a set of blueprints for the digger "which we have been asked to quote a price for making". Presumably the request was from Darby. The letter also mentions that Mr. Bentall (Baguley's chief draughtsman) "has been associated with Mr. Darby for many years and got out the whole of the

Fig. 48. A "Colchester" digger in the field
(Implement and Machinery Review 1893)
(M.E.R.L.)

drawings". No further information survives, but one wonders if this may have been a proposal for an internal combustion engined digger.

One of the last Darby diggers was inspected by King George V at Bristol show in 1913 (Fig. 49).

Darby and his sons were never to achieve the success which—if time and money spent was a guide—they so richly deserved. In the end Darby, through the Pedestrian Digger Company and the Syndicate lost in excess of £100,000, an enormous sum even now.

The reasons for failure can only be conjecture, but apart from the apparent lack of sales drive, excessive wear and tear on the mechanism, conservatism of farmers, the very appearance (particularly of broadsiders) were all factors which militated against success. In the end Darby produced a machine which would do the job it was designed to do, efficiently and—in the short term—economically. The introduction into general use of the internal combustion engine ensured the final demise of Darby's dream, and the massive Gyrotillers of 1920 to 1935 show a more than passing resemblance in design and operation to Darby's tender mounted digger (Fig. 50).

Pioneer of power "take off" Darby was in many respects ahead of his time. The story of his diggers is now ended. It is fortunate indeed that records of his endeavours have survived, even if his diggers have not.

DARBY TENDER DIGGERS 91

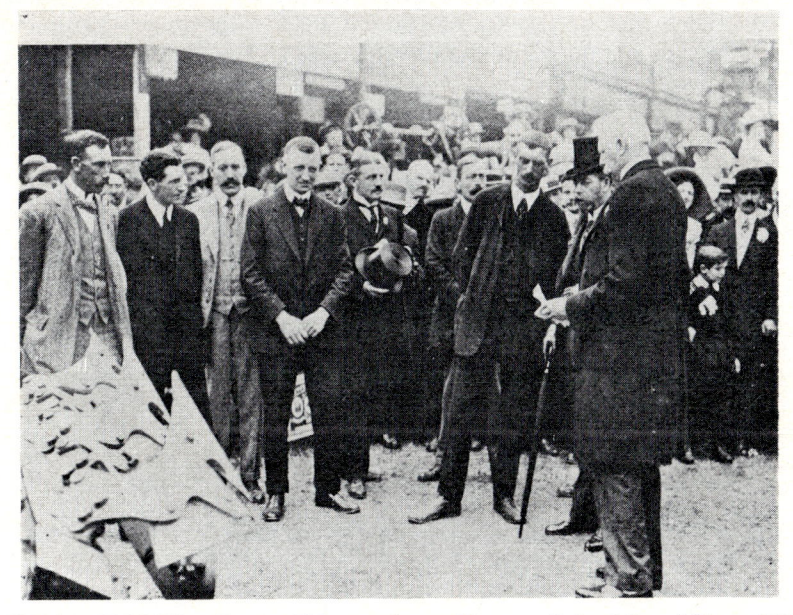

Fig. 49. H.M. King George V inspecting a digger at Bristol Royal Show 1913

Fig. 50. Fowler Gyrotiller, the internal combustion engined successor to steam diggers
(*M.E.R.L.*)

Throughout the forty years of endeavour, the known companies to be associated in the development and manufacture and sales of Darby diggers of all types were:

Agricultural and General Engineers
Aveling and Porter
Burrell
Clayton and Shuttleworth
Darby's Patent Pedestrian Broadside Digger Co.
Darby Land Digger Syndicate
Davey Paxman
W. & S. Eddington
Fowler
McLaren
Ransome, Sims and Jeffries
Ruston Proctor
Wallis and Steevens

Sidney Darby—the asthmatic junior—continued business at Wickford dealing in farm implements and spares. A countryside travelling repair service completed the facilities offered. With the closure of Stileman's Works an era passed. Steam power was of the nineteenth century, the internal combustion engine of the twentieth, fulfilment of the promise and successful application of agricultural machinery to the infant power source owed much to the labours of men like Thomas Churchman Darby.

John Mortimer wrote in 1709 some words which are perhaps appropriate to the work of Darby and others who worked to develop steam digging:

"That a higher degree of reputation is due to the discoverers of profitable arts, than to the teachers of speculative Doctrines, or to Conquerors themselves."

CHAPTER SIX

Tender Diggers—Other Types

Thomas C. Cooper

THOMAS C. COOPER of the Farmers Foundry, Great Ryburgh, Fakenham, Norfolk was responsible for establishing a Company that is still flourishing today albeit making different products from those of nearly a century ago.

In 1887 Cooper exhibited a traction engine at Newcastle (Fig. 51). Of his own design, and built for him by Garretts, it had compound cylinders 6 in. and 9 in. diameter by 11 in. stroke. Drive to the rear wheels was by means of an unusual chain made from $\frac{1}{16}$ in. thick saw blade steel. A number of engines followed from the Ryburgh works including some portable engines (Fig. 52).

Some years later, about 1894–5 a works was established on the Wisbech Rd., Kings Lynn. Cooper had been working on an idea for a digging engine for some time, and the new factory was opened for the specific purpose of producing them. It was known as the Cooper Steam Digger Company Ltd.

In 1893 Cooper made and exhibited a digger very similar in appearance to Thomas Darbys' later machines in that the digging forks were attached to the rear of the engine.

To test the principles of his digger (Fig. 53), in 1891 two sets of them were attached to modified traction engines and dug some 700 acres in Norfolk "to the great satisfaction of the owners of the land".

A complete digging engine was now produced entirely at Kings Lynn. Named "Pioneer", the engine—which was of handsome proportions—possessed three cylinders and was the first of several such engines.

Eight forks were placed in two rows across the tender set 2 ft. apart. The line of forks nearest the tender were provided with twenty tines, chisel shaped, while the rear set of forks had curved pointed tines, designed to break up the clods dug by the first row.

Engine driven cranks imparted a circular motion to the forks, which could be adjusted to dig between 3 and 8 ins. deep by lowering or raising the whole digging mechanism by means of a hydraulic cylinder which also served to raise the forks clear of the ground for road travel.

"Pioneer" had a footplate described by *The Engineer* as "a species of gallery, 9 ft. long, running across the trailing end of the boiler over the digging forks". It continues: "The coal bunker is placed on the top of the fire box, just over the fire door, and is very accessible. Indeed the whole footplate is in marked contrast with those of all other digging machines that we have seen, giving

Fig. 51. Garrett built Cooper traction engine of 1887
(*The Engineer 1888*)
(M.E.R.L.)

TENDER DIGGERS—OTHER TYPES 95

Fig. 52. A Cooper traction engine and crew provide a delightful country scene of steam thrashing about 1890

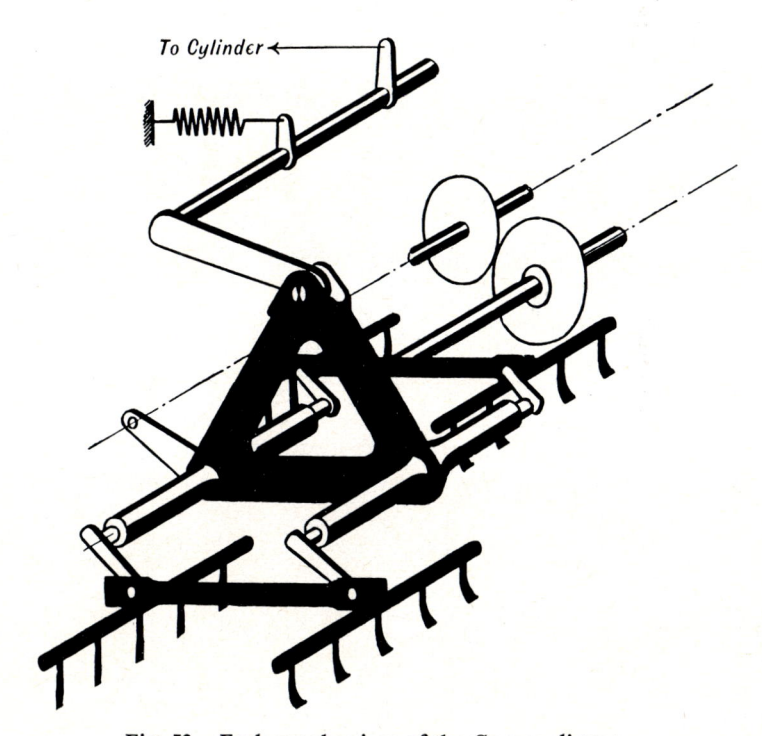

Fig. 53. Fork mechanism of the Cooper digger

plenty of room for everything. There is a steering wheel at each end of it, that is at each side of the engine, so that whether the dug land lies to the right or left of the steersman, he can stand at that side and guide the machine so as to leave no ground untouched.

"The speed at which the engine advances determines the width of spit, and there are two speeds for this; but to avoid weight and complication, instead of using clutches, Mr. Cooper carries three steel pinions, which can easily be taken off and put on the crankshaft end at the left hand side—one is for fast speed and the other for slow speed digging, and the third pinion, much larger than either of the other two, is used for travelling along the roads.

"There are three cranks on the crankshaft, which has no flywheel, and there are three cylinders placed on top of the boiler near the smokebox end, and worked triple expansion. The middle cylinder is high pressure, that on the left hand is the intermediate and that on the right the low pressure. The safety valves are loaded to 160 lb.

"The two middle forks have four tines in each; the remaining four have five tines each; they are made of cast steel—they can be taken off or put on in about a minute. These forks cost about 4s. each, and should the machine encounter a boulder, and a fork break, another one is put on within a few minutes. The depth of work is varied from 4 in. to 9 in. by fitting the points of the forks with spuds or points of steel cast in chills and as hard almost as a diamond. These save the main forks from wear. Their average consumption in Norfolk has been found to be about two per acre. They cost very little being roughly made. They are stuck on the points of the forks and secured by putting a common split pin through. Owing to the dust and dirt raised in digging, the wear and tear of the crank shaft may be very great; on one occasion for instance a journal 3 in. in diameter and 5 in. long was cut down to $1\frac{1}{2}$ in.

"Early in 1894 a twin cylinder engine appeared, and was shown at Leicester. It was described as having dirt exclusion collars at both ends of every bearing of the digging mechanism in an attempt to alleviate some of the enormous wear to which the bearings were subject.

"The boiler was fitted with a steam operated jack under the smokebox which could 'lift' the fore end to suit any level of ground."

A potential hazard with the efficiency of operation of digging engines was uneven ground. It was quite possible for the rear wheels to be on a high spot when the forks were over a low spot resulting in uneven depth of digging. Cooper overcame this problem by automatically adjusting the height of the digging crankshaft by means of a hydraulic jack which raised and lowered the forks according to the height at which a "feeler" or foot adjacent to the ground set the depth at that instant about to be dug, thus ensuring a constant depth of cultivation however uneven the surface (Fig. 54).

The R.A.S.E. awarded a special silver medal to Cooper for his engine which seems to have reached a particularly high standard of efficiency, due to the great deal of attention to detail lavished upon it.

A patent of June 1893 attempted to lighten the engine by raising the digging

Fig. 54. The handsome single cylinder Cooper digger as developed by 1894
(The Engineer Vol. 77)
(M.E.R.L.)

mechanism, allowing smaller rear wheels to be used and a lower, lighter engine.

Fowlers were involved with Coopers in the late 1890s when they attached digging mechanisms to five 10 N.H.P. compound traction engines (Figs. 55 & 56).

Later engines had a narrow track front axle, subsequently modified to a single wheel as on the later described No. 5 engine. This provided the facility of being able to turn almost in their own length, a great advantage on the headlands.

In 1900, a very handsome compound cylindered engine was produced in numbers of "several dozen" over the next ten years (Fig. 57).

The digging mechanism consisted of two rows of forks two feet apart. Six tines on each of the forward row of four forks were straight with chisel shaped ends, while the rear row had the same number of forks and tines which were curved and pointed, to break up the clods of earth lifted by the forward forks (Fig. 58 & 59). Depth of digging could be altered between three and six inches. A hydraulic cylinder lifted the whole digging mechanism clear of the ground for road travel. The engine was shown in trials at the York R.A.S.E. show at Kexby some six miles from York. The trials were supervised by Professor W. E. Dalby. A Darby Quick Speed digger was also present and the engines were performing in competition with each other.

The following figures provide a comparison between the contenders, the

Fig. 55. Fowler built Cooper digger of 1899
(*M.E.R.L.*)

Fig. 56. Rear view of the Fowler-Cooper digger. The double steering chain enabled a clear view from both sides of the engine
(*M.E.R.L.*)

Fig. 57. Cooper digger of 1900. Probably the first type to be completely constructed by Coopers in quantity

Fig. 58. A narrow front track Cooper digger with forks flying
(*Implement and Machinery Review* 1900)
(*M.E.R.L.*)

Fig. 59. Drivers eye view of the twin cylinder Cooper digger. Note the horizontal governor and neat layout of the engine
(*The Engineer 1896*)
(*M.E.R.L.*)

Cooper machine being superior in performance according to the judge in awarding first prize for its lightness, superior work and economy.

Cooper engine
    Weight of engine and digger in working order 11 ton 17 cwt. 2 qr.
    Weight of digging mechanism 1 ton 17 cwt. 2 qr.
    Front wheels 3 ft. 6 in. dia. by 12 in. wide.
    Rear wheels 5 ft. 9 in. dia. by 26 in. wide. These could be reduced to 20 in. wide for the road.
    Cost £750 complete compound engine $6\frac{1}{2}$ in. and $11\frac{1}{2}$ in. bore × 12 in. stroke. Working pressure 150 p.s.i.

Darby engine
    Weight of engine and digger in working order 12 ton 10 cwt.
    Weight of digging mechanism 2 ton 10 cwt.
    Front wheels 4 ft. dia. by 9 in. wide.
    Rear wheels 6 ft. 3 in. dia. by 18 in. wide.
    Cost engine only £845, digger £350. Simple engine.

A larger 12 N.H.P. engine was produced soon after 1900 with compound cylinders 8 in. and 13 in. × 12 in., working pressure 120 p.s.i. Cylinders and motion were now mounted on a platform positioned on top of the watertank and behind the manstand. Two $\frac{1}{2}$ in. hornplates were riveted to the raised top firebox of the stubby boiler. Eight and four forked digging mechanisms could be attached (Figs. 60 & 61).

Not all the engines were sold directly to customers, for some were kept standing by at the works, complete with crews for hiring out to farmers in East Anglia, although they rarely went beyond the borders of Norfolk and Suffolk.

An anecdote regarding one of the hired engines concerned a farmer who required his land to be worked in the vicinity of Setch bridge. The wheels were of narrow tread on this particular engine and when driven over the bridge to the site, was found to sink badly into the rain softened soil. After much labour in extricating the engine, it was sent back to the works to have wider wheels fitted. It was only when the engine had travelled all the way back to the bridge that it was found to be too wide to cross!

A third type of digger classed as the No. 5 was produced in 1903 (Figs. 62 and 63).

This was a small machine which was supplied as a basic traction engine to which could be attached several types of mechanism.

With only one front wheel, the steering wheel was also at the front and to one side in a reversion to earlier Aveling-type practice, the steersman sitting almost over the wheel, being able to steer the engine round in almost its own length. The cast rear wheels had square strakes.

Compound cylinders $4\frac{1}{4}$ in. and 8 in. by 9 in. worked on a pressure of 180 p.s.i. from a quick steaming, return flue boiler giving the engine a 6 m.p.h. turn of speed. The cranks, connecting rods and valve gear all worked in an oil bath and were totally enclosed, and the engine was fitted with Pickering governors.

As with its larger brothers the little No. 5 engines attachments were at the rear and took the form of four three tined forks digging a normal width of 5 ft. 10 in. which could be increased to 6 ft. 10 in. for very light soil, giving a digging rate of between 4 and 6 acres per day. Maximum depth dug was 9 in. and instead of hydraulic lowering and raising of the forks, hand levers were provided.

Other attachments were available in the form of cable ploughing apparatus. Either single or double drums mounted at the rear were parallel to the hind axle in a similar fashion to the Howard Farmers Friend ploughing engine.

When equipped with single drums, the engines were supplied in pairs, and those for roundabout ploughing singly, with double drums. An automatically contracting external brake band prevented over-running of the drums, which were each supplied with several hundred yards of rope.

Coopers did not seem to supply the implements to go with the cable ploughing engines, presumably leaving this to the selection of the customer, but it was claimed that a depth of 20 in. could be achieved.

Most Cooper diggers were exported, twenty four going to Egypt on one order alone (Figs. 64 and 65).

Some work was done in the form of a drawing on a further small digger—

Fig. 60. Cooper undertype engined digger, typifying the smooth and uncluttered designs of this Company
(*Cooper Roller Bearings Co. Ltd.*)

Fig. 61. Rear view of the compound undertype digger of 1900
(*Cooper Roller Bearings Co. Ltd.*)

TENDER DIGGERS—OTHER TYPES 103

Fig. 62. Cooper No. 5 engine of 1903 equipped for digging. Alternative attachments included steam ploughing cable gear

Fig. 63. Rear view of the little No. 5 digger

Fig. 64. Cooper chain driven digger at work, probably in Egypt

Fig. 65. Cooper chain driven digger being erected in Egypt

the No. 6 type, and it was very similar to the No. 5. Whether it was actually made is not known, but as with other steam engine makers at this time, Coopers were probably persuaded it was of no avail to develop diggers further, in the face of competition from internal combustion powered tractors which by this time were becoming widely used, although it is something of a mystery why Coopers did not try to adapt their diggers to the new power.

In 1907 the Company commenced manufacture of roller bearings, for which the Cooper Roller Bearings Co. Ltd., are today renowned. It was indirectly due to diggers that this came about, when the engines on hire were required to travel long distances, they hauled three trailers. One loaded with spares, one living van and one carrying coal and water. As time went by, the trailers became more heavily loaded and the engines experienced difficulty in hauling the extra load. To reduce friction in the wheel hubs, they were bored out to a larger diameter and short lengths of mild steel rod inserted between the hub and axle end. These were the first roller bearings and formed the basis of the new business after production of digging engines ceased, thus making Coopers the only Company who can claim to be founded directly from the manufacture of digging engines, also commercially speaking the most successful of the many who attempted to till the land by means other than the plough.

John D. Garrett

A model of a new type of digger was shown at Kilburn in 1879. Designed by John D. Garrett of Southwold, Suffolk, formerly of Buckau, Magdeburg. The engine was intended to work on hard stubble land.

Early in 1883 an engine was constructed. It had twelve sets of forks arranged in line across the rear of the engine at right angles to the direction of travel. To assist in breaking up the clods, disc coulters were placed between each set of forks, and to invert the soil, the forks were twisted sideways at the end of their digging motion.

Built at Buckau, the engine was chain driven and worked successfully for two seasons in the Magdeburg district, cultivating to a depth of 24 in. at the rate of 2/3rd. of an acre per hour for which 18s. (90 p) per acre was "willingly paid".

The second engine (Fig. 66), which does not seem to have reached constructional stage was modified in having only ten sets of forks. The twisting motion of the forks and the disc coulters were now eliminated.

Two forward speeds were provided, one of 2 m.p.h. and a cultivating speed of 48 ft. 3 in. per minute or just over $\frac{1}{2}$ m.p.h.

No more is heard of Garretts digger, a rather overbearing article in *Engineering* commenting in 1879 with reference to both Garrett's and Darby's diggers: "We must own to being opposed to any system of steam cultivation which involves the continuous transit over the land of a heavy engine and its attachments' , a hearty condemnation of an idea before being given a fair trial and an example of someones personal prejudice showing through in an article which

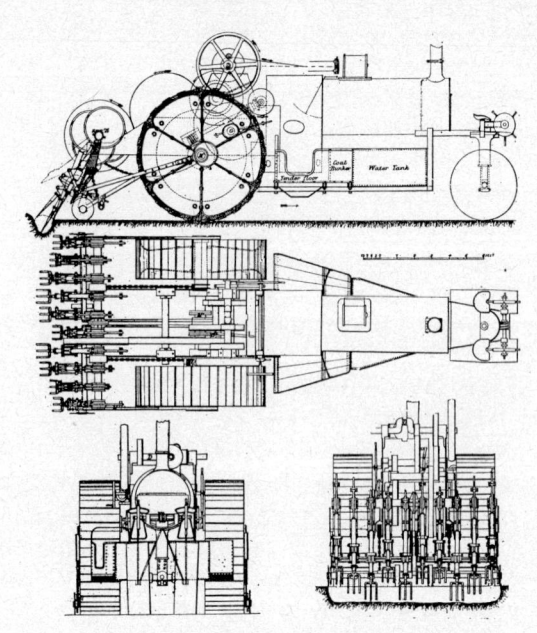

Fig. 66. John D. Garretts' digging engine of c. 1882. The narrow track front axle is steered from two positions over the axle, while ten three tined forks performed the work at the rear
(*Engineering 1883*)

should have devoted itself to more pragmatic matters instead of bucolic pretence.

D. Nagy

In 1886, D. Nagy of Budapest patented a digging engine which in many respects was similar to Calloway and Purkis' engine of 1849 (Fig. 67). Arranged across the rear of a traction engine, two chains with cultivators attached revolved around the periphery of the two arms which projected out from the engine, cultivating a swathe behind it as it progressed.

J. Johns

J. Johns of Ipswich designed a digging engine in 1889 which should in theory have produced the ideal seed bed, complete with sown seed. Not satisfied with utilising a single method of cultivation, Johns patented the only digger to incorporate, disc coulters, forks, vertical rotary cultivators, seed drills and a horizontal rotary harrow, all mounted in a huge frame attached to the rear of a comparatively small engine. This project was far too ambitious and no more is heard of it after the patent specification.

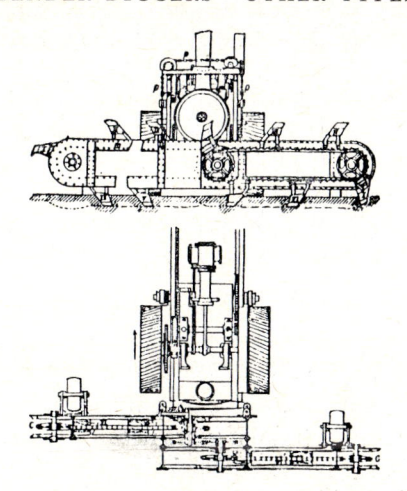

Fig. 67. Nagy's digger of 1886

Boghos Pacha Nubar

An Egyptian representative of the art of digging by machine was one B. P. Nubar, who in 1897 proposed a machine which was to be propelled by the—as yet—infant "auto motor" or internal combustion engine, with steam as an alternative. Whether this machine was in fact built is unknown, but it must rank among the first proposals for a tractor-like machine, heralding the demise of steam in general and steam ploughing and digging in particular (Fig. 68).

The tractor itself was equipped with rotating ploughs or screws arranged transversely underneath the chassis. These were intended to bore their way through the soil, thus breaking it up. Similarities could be seen in the principle

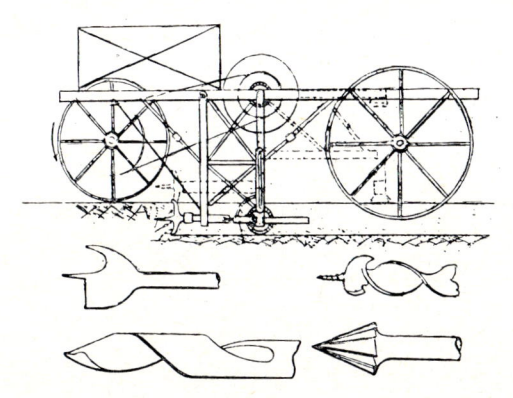

Fig. 68. One of the first agricultural machines to be powered by an internal combustion engine, Nubar's tractor like cultivator of 1897, cultivating by means of a screw plunged into the soil

to Garroods rotary auger engine of 1892 and may have provided inspiration to Nubar.

Nubar had previously designed the rotary cultivating engine shown in Fig. 8, about 1890, but little is known of this enterprise other than that the engine was possibly of French make.

Charles Garrood

Charles Garrood of South Lodge, Forest Hill proposed a novel type of rotary cultivator in 1892 (Fig. 69). It consisted of a series of contra rotating screw-like cutters, intended to break up the soil into a fine tilth by literally drilling their way along the soil. A frame at the rear of the engine carried the "convoluted" screws also the train of gears necessary to drive them. Belt drive from the flywheel of the engine enabled the whole frame to be easily dismantled from the engine, which could then be used for any other purpose.

Almost certainly this was one of the ideas that never left the drawing board. It is difficult to see how this type of digger could compete with established tender mounted forked diggers of Darby, Cooper and Proctor, which by 1892 had reached an advanced stage of development.

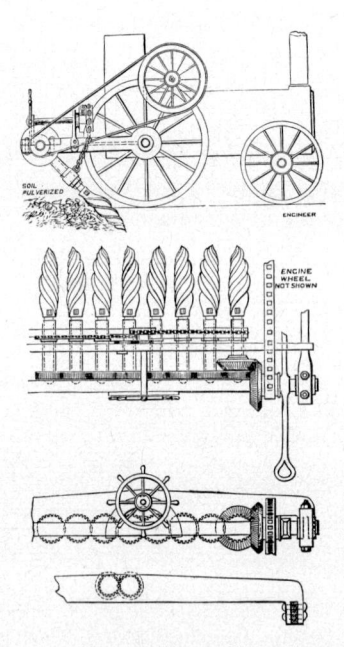

Fig. 69. Charles Garrood's rotary cultivator. A train of screw like augers pulverised the soil
(*The Engineer 1892*)
(*M.E.R.L.*)

Frank Proctor

Frank Proctors digging engine was one which was to prove quite successful and was made mainly for export. How many were built is unknown and few performance details survive, but it was made in both 4 H.P. single cylinder and 7 H.P. compound versions. The illustration shows a neat, diminutive engine, if the driver is in correct proportion to it. The horse power quoted would seem to bear this out and one wonders if the weight of about 4 tons would have been great enough to ensure good penetration of the forks without lifting the engine bodily in certain conditions (Fig. 70 and 71).

Constructed by Burrells, the first engine was delivered on 14 August 1886 for a nominal £350, presumably to enable its performance to be evaluated. A compound engine went to the Marquis de la Laguna, Spain and the engine illustrated here was sent to a sugar refinery at Waghäusel, Germany. It was demonstrated on 4 December 1886 to notables, among whom were the Grand Duke of Baden and representatives from France, Hungary and Russia. "It met with so much success", reported a delighted Proctor, "that more orders are now in hand than can be executed in time for this years spring and autumn ploughing, or rather diggings."

This engine was probably the only digger made by Burrells, their faith in steam ploughing being that much greater, and as it was made to the order of

Fig. 70. Frank Proctor's Burrell built digger. While only of 4 h.p. the driver appears to be something of a giant!
(*The Engineer 1886*)
(*M.E.R.L.*)

Fig. 71. Rear view of the Proctor digger which appears to be a larger engine, perhaps the 7 h.p. version. The alternative is that the driver is a dwarf!
(*M.E.R.L.*)

Proctor, they were saved the expense and time of developing the engine, reducing their financial risk to the manufacture of the engine.

Proctors engine was awarded a medal by the Essex Agricultural Society when it was shown at Ilford in 1888 and there was mention of the possibility of a steam digging syndicate to promote the engine in 1891.

The engine itself had three sets of forks mounted in an extension rearwards of the hornplates—replacing the traditional tender—and operated by a triple throw crankshaft which was driven by a train of gears from the engine crankshaft. This gave a digging motion opposite to that of Darbys mechanism, as Proctor went to great lengths to point out.

A joint patent for the complete digging mechanism was obtained by Proctor and Burrells in 1884 (Fig. 72).

It was important for safety reasons that the engine could perform only one function at a time; that is either be in gear for road travel or in another gear for digging and inching along. A clever combination gear ensured that the two could never be used simultaneously and allowed the digging and inching drives to be in or out of gear together but not individually.

For several weeks the pages of *Engineering* contained a curious discussion between Proctor and that other exponent of digging John Knight.

A short article appeared on 26th September 1890 on the details of the forks of Proctors engine in which he illustrated the working mechanism.

Next week Knight wrote to the editor "Will you allow me to say that this motion (of Proctors forks) is identical with that used by me fifteen or sixteen years ago, and applied in digging machines for hop gardens? Several of these machines were made and used, not only in the Farnham hop district but in Kent...."

Proctors rather patronising reply included the passage: "the nature of the work (of Knights machine) satisfied practical judges, which shows that the *action* of the forks was a correct one; whether it was identical to mine I cannot

TENDER DIGGERS—OTHER TYPES 111

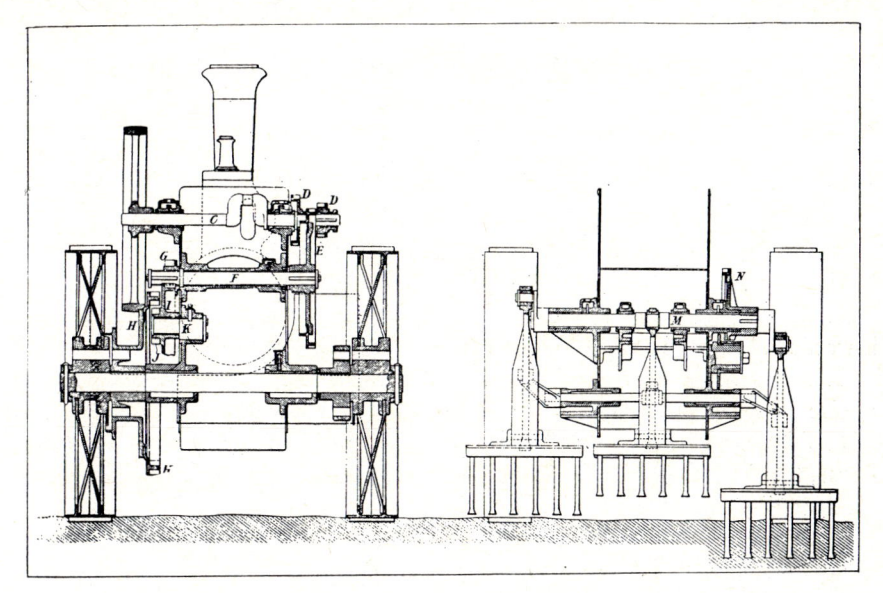

Fig. 72. This specification drawing of the Burrell-Proctor single cylinder digger gives no clue as to its actual size
(*Patent Office*)

tell as I have never seen one of his machines and have no drawing of such by me, but I will take Mr. Knights word for it that it was so.

"I do not suppose that Mr. Knights machine could be used for general cultivation work, because in digging up hard dry land (which is a most valuable condition for it to be in for cleaning and draining land) the power must be transmitted direct on to the forks from the engine, and *not* through the medium of a flying rope. Nevertheless, Mr. Knight has undoubtedly earned the distinction of being one of the early pioneers in the system of steam digging, which no one who is now engaged in this occupation should grudge him".

Satisfied that honours had been duly given, Proctor continued developing his digger, Knight replied without acknowledging the magnanimous gesture: "Mr. Proctor is quite right in saying that my digging machine was not suited for hard dry land. It was schemed at a time when labour for digging hop grounds was scarce". That was the end of the matter.

Rather cannily, in his correspondence with Knight, Proctor omitted to mention his own words written to that other great engineering journal *The Engineer* three years previously when he acknowledged that his digging engine fork mechanism was in fact the same as M. R. Pryors. Proctor seems to have attracted accusation of "idea poaching" for some reason, as in this instance he was defending his mechanism against Darbys . . . "the movement of the forks of my digger is exactly opposite to the movement of Darby's, Parker's, or any other inventors. I make mine on the lines of Mr. M. R. Pryors patent of 4th April 1884, whereby the digging shaft revolves in the opposite direction, and operates at the top of the fork handles, the resistance being given below instead of *vice versa* as in former ones" (Fig. 73).

An interesting paragraph in Proctors original letter of 1890 describes the alternative methods for digging by steam.

He stated that it was not possible to modify a standard traction engine—"owing to the shaft centres being unsuitably arranged the work has proved unsatisfactory". He thus advocated a purpose built engine.

The three methods of digging with tender mounted digging apparatus such as his, were:

1. The machine travelled round the outside of the area being dug, working inwards "similar to the path usually taken with the sheaf-binding harvesters".

2. Work was commenced in the centre of the field and described "a number of circularly ended rectangles" until the whole area was dug.

3. Digging in straight lines up and down the field as if ploughing with a horse, turning at the headlands by "reversing the engine much as is done with steam engines in reversing them by triangles".

This last method claimed Proctor, was the most commonly used, turning at the headlands taking only a little longer than with a team of horses.

J. A. Clarke

An intriguing drawing, in the possession of the author shows a digging engine with some novel features. The general arrangement drawing—which is

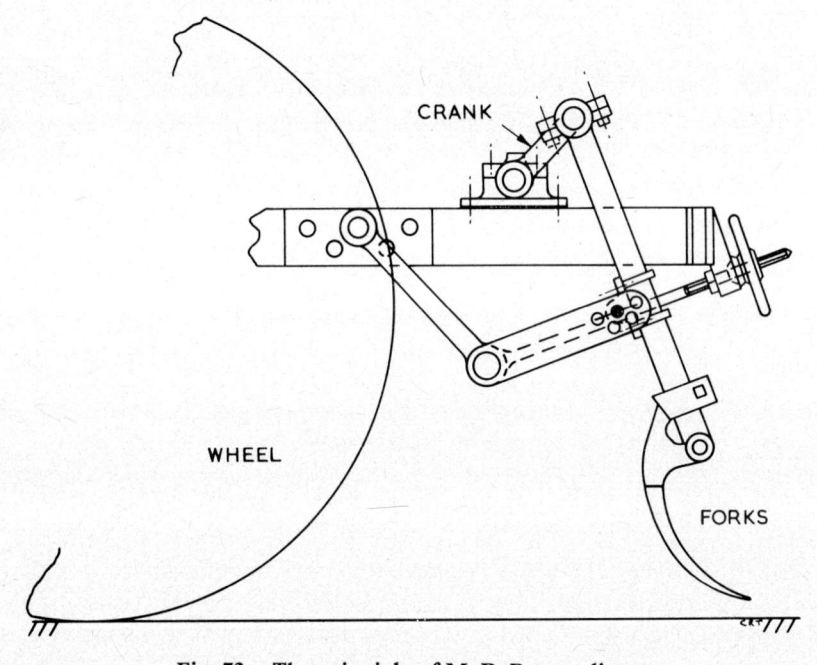

Fig. 73. The principle of M. R. Pryors digger

undated—is entitled "J. A. Clarkes ploughing engine" shows a three wheeled engine of small size. Utilising one of Frederick Savages twin cylinder centre engines and boilers, more familiarly powering roundabouts, it may be that this was intended to be a small prototype digger to try out the principle of Clarkes ideas.

The rear wheels were positioned astride a device consisting of two chain wheels one of which was on the same axle centre as the rear wheels and the other some distance forward, very similar in principle to Major Pratts design of 1810. A continuous chain passing round the wheels possessed pivoted sets of three spikes eight in number, which by means of a cam, were made to project out only when in contact with and digging into the soil, folding up again at the end of travel, passing along underneath the engine and then unfolding for the digger stroke, and so on. Almost certainly never built, the drawing may be the only surviving evidence of practical work done by John Algernon Clarke of R.A.S.E. lecture fame and a known digging enthusiast. Chain drive to the rear wheels suggests a date in the late 1860s as it was at this time that Savages were producing their chain driven traction engines.

Unfortunately, records of engines made by Savages before 1870 have disappeared, and it is therefore uncertain how far this little digger progressed, but as there is only the one general arrangement drawing, it can be fairly safely assumed that it remained an interesting but still-born idea.

Another digging engine drawing in Savages records possessed two cylinders mounted one on each side of and at an angle to the boiler, in much the same layout as Stephensons renowned "Rocket". Contemporary with Clarkes engine, it too remained a paper idea.

Fig. 74. John Henry Knight's hop digging machine

CHAPTER SEVEN

Hop Diggers

John Henry Knight

JOHN HENRY KNIGHT of Weybourne House, Barfield, Farnham was born in 1847. He was a hop grower and engineer. The digging machine that he commenced developing in 1872 was notable in that it was specifically made for working in hopfields, notoriously difficult to cultivate with economy and efficiency.

In his ultimately unsuccessful attempt to solve this problem, Knight considered that a machine replacing expensive and sometimes unreliable manual labour would be best designed to imitate the human action of digging.

The machine was made to imitate in ingenious fashion, the actions of large numbers of labourers required to dig the hop fields, and of course to reduce their numbers, if not completely eliminating them.

The *Agricultural Gazette* of 1876 published an article on the progress of hop diggers up to that time, including some interesting remarks regarding others than Knight who had tried to develop machines to do the same job. They are mentioned fleetingly as failures, but now the only surviving information on otherwise forgotten enterprises which—to the inventors—were no less important as any digging machines of which there is more detailed information to record here, and which although better publicised were in the end no more successful.

The *Agricultural Gazette* said:

"Hope growers have long felt the want of a machine to dig hops to supersede labour. This work is very arduous, and prematurely bends the backs of the labourers employed in the hop districts. It is also an expensive process, costing from 19s. to £1 4s. per acre, according to the state of the soil and the supply of labour. Since the agitation among the labourers prices for digging have risen considerably, and in some districts it has been difficult to get the hops dug at all. Men do not like digging if they can get lighter work, especially after they are middle-aged. The tendency also is to scamp the work, merely to push great spits of earth over. It is very rare in these days to find a man who fairly lifts the soil and turns it over, who buries the manure and weeds in a thoroughly workmanlike manner, after the fashion of his less enlightened, but more ingenious forefather. Some hop-growers, in despair at the slovenliness and expense of modern hand digging, have adopted ploughing their hops with small turn-wrist ploughs, drawn by two horses; having men at the same time digging the slips and stack places. This is by no means a satisfactory process; but it appears to suit the hops almost as well as digging, and it renders the growers more independent with regards to their labourers. A digging machine was shown at the Royal Agricultural Society's Show at Salisbury, whose principle, it was thought, might be adapted to hop cultivation. This was designed according to Talpa's theory as to the action of a moles feet upon the soil, and was merely a revolving cylinder garnished with spikes made broad and clam like at their extremities. This was soon consigned to the limbo of machines whose mission has failed.

"At Wolverhampton show, prizes were offered by the Royal Agricultural Society for an implement for digging hops. Though this prize was not awarded, no machine being entered that possessed a particle of merit or of novelty, the offer of it set men thinking about diggers and digging.

"The next year at Hull, a digging machine was shown which was supposed to be suitable for hop land. As it was too wide for the alleys, its unsuitability was obvious to anyone who had ever seen hops growing; still, it had some noteworthy points of improvement upon its Salisbury prototype.

"Mr. Knight of Farnham, has recently patented a digging machine specially for hop land, which has some very clever and novel points, and is a most ingenious and efficient substitute for digging by hand. This machine is driven by an ordinary portable steam engine. It is not drawn or dragged as other cultivators and diggers, but is propelled by two large hind wheels driven by an upright shaft set in motion by a grooved horizontal driving wheel connected with the engine by a high speed cord running on pulleys. There are two smaller wheels in front to take part of the weight of the machine and for steerage purposes. Upon a crank shaft of 'three throws' connected with the driving shaft by a bevel wheel gearing into a pinion, three vertical rods are fastened, which are kept in their places and are guided by other rods or guides. These vertical rods have each a frame at its extremity into which three or four tines are fastened, just like the digging spuds used in the hop plantation, only that the tines or 'speens' are sharper and not so much flattened out at their bases.

"When the machine is in motion (Fig. 75) these forks or spuds are forced into the soil evenly and regularly, being guided uniformly by the guide rods,

Fig. 75. Knight's hop digger at work. This is one of the Howard built machines (*M.E.R.L.*)

taking 'spits' or furrows of about five inches wide, lifting the earth and throwing it, not exactly over, perhaps with perfect accuracy, but moving it, disintegrating it far more than a hand digger could, and fairly burying the weeds and manure. The movement or action of the forks is remarkably good—a wonderful imitation of the action of the human arm using a spade—and the greater the speed the better is the work done; the small clods and dust being literally tossed into the air. One man steers the machine down the hop alleys, and another is required behind to throw the driving shaft in or out of gear, and to fetter the inner wheel while the machine is turning, which is accomplished by a clutch fixed to the axle. This machine has been doing good work in the Farnham hop district, and was recently exhibited in Kent, near Maidstone, upon hop land in the occupation of Mr. Charles Chambers, at Langley. Mr. Knight superintended the working of the machine, which was driven by an 8 horse portable engine belonging to Messrs. Garrett & Co. Owing to the dry weather the land was in capital order for the trials, and there were not many weeds; it had been manured with fur waste, or 'rabbit clippings'. The machine moved the soil to an average depth of $9\frac{1}{2}$ inches, with an average width of spit of about $5\frac{1}{4}$ inches. All the soil was moved, though it was not all turned over, while the weeds and fur waste—at all times difficult to bury—were as well covered as most of the labourers cover them. Many of the leading Mid Kent hop growers witnessed the trials of the machine, and expressed much astonishment at its performance, and were thoroughly satisfied with the work done. Mr. Knight believes that it will dig from $3\frac{1}{2}$ to 4 acres per day, at a cost to a grower who buys a machine, costing from £125 to £145, and a steam engine to drive it, of 12s. to 15s. per acre including wear and tear, coals labour, and interest on outlay. It is thought this is much under the mark, and that it would cost much more per acre. The expense per acre must depend upon the extent of a purchasers acreage; but even if it cost 22s. per acre to dig hops in this way it would be a great swing ultimately. A large extent might be dug in a few weeks, and the staff of men who had to be kept on for digging, and are often kept to work at unprofitable jobs in frost or snow, and in wet weather, when they cannot dig, might be much diminished. Probably the persons who let out steam ploughs and thrashing machines in the hop districts, will also let out these digging machines to their own profit and the advantage of the growers, the greater part of whom have only from 25 to 40 acres, whom it would not pay—who could not afford to buy a machine and an engine. No doubt this hop digging machine will be still further improved, and alterations made in its details. Its principle is correct, and much credit is due to Mr. Knight, its inventor and patentee, for having embodied it in such a practical form. As it can be applied to digging of all kinds of land it will be a most useful agricultural implement which deserves encouragement. And now that a *bona fide* digger has been invented at last, the Council of the Royal Agricultural Society should renew the prizes for the best hop digging and the best general digging machines in the year 1878."

Knight describes the use of his machine in a letter to *Engineering* in 1890.

"All your readers may not know that hop land is dug in winter or early spring, when it is often very soft, and would be quite unable to bear the weight of a digging machine carrying its own motive power, hence the use of a fly rope. The weight of the diggers was about 30 cwt. and this was quite heavy enough for digging (say) in January or early in February, the machines often having 'stuck in the mud' when crossing wet parts of the field. In dry weather and over hard ground the machine was weighted to prevent it lifting and riding on the digging forks; the speed of the crankshaft (driving three forks) was 120 to 150 r.p.m. One great advantage of this digging machine was that any engine of 8 H.P. and upwards could be used for driving it. All that was necessary was to key a grooved pulley on the crankshaft to drive the high speed cord."

Perhaps the best way to illustrate Knights attempt to perfect his digger is to use his own words as written in a note on his "Engineering Reminiscences".

"About this time there was a rise in wages of agricultural labourers and men were hard to get. I could not get my hop-ground (at Badshot Farm) dug till long after it should have been, I therefore schemed a digging machine, a model of which I made just before going to Egypt. But this model was very crude; nevertheless it dug and covered over very fairly well bran thrown over the surface of the ground to represent manure (this was a $\frac{1}{2}$ scale machine).

"On return from Egypt I had a working machine made by Hetherington and Parker of Alton (Figs. 76 & 77). They were so long in making the machine that it was quite late in the autumn before it was tried. It then required alterations and there were many breakdowns, so that it was not of much use for hopground digging that season.

Fig. 76. A Hetherington and Parker digger. Possibly the first one to be made for Knight, this original works photograph is stamped "Wey Iron Works, Alton, Hants, 13th December 1876. The hopper like box contained stones to add weight (*M.E.R.L.*)

Fig. 77. Rear view of the Hetherington and Parker digger (*M.E.R.L.*)

"It was worked by a flying rope driven from the grooved rim of the flywheel of a portable engine, so that there was no motive power on the digger but only the digging machinery; the object of this was that the machine should be light enough to pass over soft ground without sinking.

"The difficulties were very great but by about the end of the second season I had a machine that worked fairly well. There was still some delay through breakdowns; there was sometimes trouble in turning at the ends, for the digger propelled itself as a traction engine does and to turn, one wheel was put out of gear, and the other turned it round; the forks were not properly proportioned so that the ground was not completely turned.

"In the spring of 1875 I had it on view at Badshot and many farmers came from various parts, Kent etc. to see it. In the autumn I had it 'working' on view at Langley Park Farm near Maidstone and a great many came to see it. The farmer C. Chambers gave a spread but there were one or two most unlucky breakdowns.

"So before another season I had another machine made all of iron by Howards of Bedford (Figs. 78 & 79). This worked very well and I took it to Canterbury and in the spring I took a contract for digging some sixty acres near Ashford, hiring a traction engine and got on fairly well.

"I showed it at Bath Show (B. & W. E.) in 1877 and year following at Bristol Royal Show.

"I dug my own hopground both at Badshot and Dippenhall (a second farm to west of Farnham) with it for some years averaging 4 acres per day. At last I sold the traction engine and the digger was laid aside.

"I spent a great deal of money on the digger and got no return. Now there are diggers made by Proctor and Darby; these are self contained and not suitable for hop grounds.

"I made a good many acquaintances and friends over the digger and altho I lost heavily over it, I do not altogether regret it. One cause of failure was in the first machine's inferior work causing endless breakdowns, but the system could never have answered—the wear and tear of the fly rope was too great."

There is some doubt as to the number of machines constructed. Knight's own account mentions only two full size machines, but "Iron" mentioned a number four machine built by Howards to the pattern of machine number two.

An examination of pictorial evidence throws some light on this point.

Of Knight's $\frac{1}{2}$ scale model nothing is known, except of its existence and can be disregarded as a working machine, being made merely to demonstrate working principles.

Study of the surviving photographs show that four full size diggers were made and—thanks to the remark in *Iron*—the sequence in which they were made.

The digger shown in Figs. 76 and 77 is the first one, with which Knight had so much delay in delivery and trouble in getting to work. The original sepia photographic print from which the picture here was obtained is marked with Hetherington and Parkers works stamp on the back. It states that their factory

Fig. 78. Howard's version of Knight's digger appears to be a heavier machine which dispenses with the stone box of earlier diggers. Made in 1876
(*M.E.R.L.*)

Fig. 79. A side view of Howard's digger
(*M.E.R.L.*)

was named Wey Iron Works, Alton, Hants and the date 13th December 1876. According to Knight his first machine was made by the autumn of 1874, so it must be assumed that the photograph was printed some two years later, when Howards had taken over the construction of Knight's diggers.

By the autumn of 1876 Knight had *"a machine"* (not *the* machine) working, suggesting that another one had by this time been made. Photographs certainly show another digger not made by Howards. In contrast to the first digger which bears Hetherington and Parker's nameplates, Figs. 80 & 81 show no indication of the maker, but similarities in design and construction suggest that the digger shown was probably also made by Hetherington and Parker. Closer inspection of these two photographs show certain slight differences in detail between them—for example, a chain appears on the front axle of one but not the other, but these are probably modifications made between the photographing of the digger to improve performance, although the possibility of a third digger cannot be ruled out.

So far, therefore, we have two with the slight possibility of three Alton built diggers.

In 1876 Howards of Bedford made their first digger for Knight, almost certainly the one shown in Figs. 78 and 79. They went on to build another machine (Figs. 82 & 83) which, it will be noted is very similar to the one in Fig. 80 and therefore must be the fourth digger to the pattern of machine number two, as mentioned in the valuable clue in *Iron*.

Thus, it can be deduced that if the total number of diggers made were four —as stated in *Iron*—that the sequence of building was Figs. 76, 80, 78 & 82. Add the original half scale model and the sum total of Knight's diggers is apparent.

Movement of the fork tips, when compared with Thomas Darby's, was crude. Knight's forks were mounted on the end of long shafts or handles. Three forks were incorporated on all machines, but the number of tines varied from 4-4-4 on the first machine, 4-3-4 on the second to 5-4-5 on both of Howard's. All tines were straight and in line with the handle. Mechanical motion was imparted to the fork tips by a simple three throw crankshaft which plunged the tines at an angle into the soil (Fig. 84). A horizontal arm acting as a pivot for the handle then pushed the tines in an upward arc, lifting the soil with it. At the top of the arc the tips moved horizontally backwards, then repeating the motion. One would have thought that a curve forward on the tines would have not only lifted the soil, but helped to turn it over as well, but Knight did not appear to criticize his machine's ability to dig, finally terminating his project on the grounds of excessive rope wear. Depth of digging was altered by a screw adjustment clearly shown in Fig. 79, which altered the entry angle of the forks, thus increasing or decreasing the depth of penetration.

All diggers, with the exception of the first one which had worm wheel steering, traction engine style, were guided by a steersman who operated a tiller like handle to the fore end of the machine. The third digger (Howard's first) possessed a single front wheel, the others having two wheels and axle,

122   DIGGING BY STEAM

Fig. 80. This digger may have been constructed by Hetherington and Parker and appears to incorporate a stone box
(*M.E.R.L.*)

Fig. 81. Knights digger in use. This photograph is the original of the engraving at the head of this chapter. If the reader wishes, an interesting five minutes can be spent in spotting the differences!
(*M.E.R.L.*)

HOP DIGGERS

Fig. 82. Howard's second version of Knights digger. Now of lighter construction. Note the handwheel which adjusts the depth of dig
(*M.E.R.L.*)

Fig. 83. A half rear view of the second type of Howard's digger
(*M.E.R.L.*)

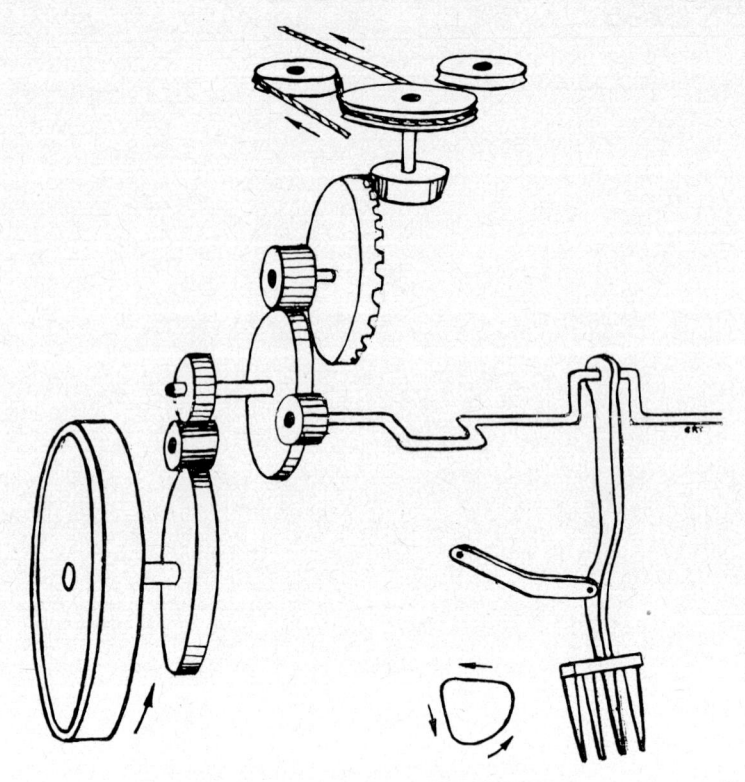

Fig. 84. The principle of Knight's digger
(Willmer House Museum, Alton)

probably providing better grip and steerage when turning the machine in its own length between the hop alleys.

A box within the framework of all except the fourth digger contained ballast which helped to increase adhesion of the two larger wheels and provide positive steerage. Additionally, "Spuds" were attached to the rims of the larger wheels while the digger was working.

The digger was put into operation by means of a "flying rope" which passed around three pulleys mounted on top of the machine, their centres arranged in triangular planform. No photographs seem to have survived of the digger and attendant engine together. Both traction and portable engines could provide the necessary power via a hemp rope which passed around the area to be dug, carried through snatch blocks and rope porters, in similar fashion to Fiskens roundabout ploughing system. The brothers Fisken utilised two self moving anchors between which the balance plough travelled to and fro. Knights digger was similarly driven, but being a one way implement, it was necessary to turn it at the end of each traverse. Instead of anchors, weighted carts with pulleys attached provided anchorage and these were moved manually along to keep up with the digger as it worked its way across the hop field.

Knight describes his method of keeping the hemp rope taut, while the digger was being moved along:

"The driving cord is led on to and off the driving pulley by guide pulleys so arranged that the digging machine will work either to or from the engine, or even at right angles with the line of ropes without reversing the engine or changing the position of the ropes on the pulleys, so that the engine has always a direct pull on the digger (Fig. 85). A wagon on the headland opposite the engine carries two or more pulleys on vertical spindles, which turn the cord at right angles, that is to say, towards the tension pulley. This tension pulley is fixed at one end of a sledge, on the other end of which is a weighted box to balance the pull of the rope when the wagon is hauled forward. The tension pulley on the sledge follows, and by its resistance over the ground keeps the rope tight" (Fig. 86).

When the digger had reached the headland, the forks were lifted by means of a handwheel (or crank) which depressed one arm of the bell crank on the other end of which the radius rods were attached. As it was forced back the forks lifted out of the ground, when the land side wheel was released by turning the hub mounted handles, and withdrawing the dog clutch clearly seen in Fig. 77.

The engine and tackle were operated by a crew of three men and a boy, one engine driver, one to operate the clutches on the digger which was steered by the boy, and one to move the anchor carts.

The similarity of operation between Knight's and Fisken's systems is more than coincidence. Fisken's rope speed of 22 m.p.h. was however, far exceeded by Knight's rope which travelled at an amazing 34 m.p.h. with a considerable risk of danger of decapitation to the operator, and no doubt the cause of the excessive rope wear. Rope breakage was minimised by means of a weighted arm which allowed movement on one of the pulleys should the rope snatch for any reason.

The rope was driven from a vee groove in the engines flywheel.

Knight obtained two patents for his digger. The first, of 1873 describes a machine which is similar in principle to the Hetherington and Parker machine, but differs considerably in detail—a feature of patents which understandably, tend to reflect the ideas of the inventor, rather than the practical assemblage of gears, pulleys, nuts and bolts which emerges from the factory door to be tested in action, incorporating the draughtsman's schemes designed to make the patent design a mechanical actuality rather than a theoretical possibility. If Knight's original plans were as on the patent, it is quite likely that this was the reason for the long delay in delivering the first machine, but this could be unfair comment! An additional idea for making use of the digger was to remove the forks and attach a cultivator, harrow, or even a plough, when the machine "acts as a motor only", but it was not put into practice, digging being the prime purpose.

The second patent of 1876 described two modifications to his original design. To enable the digger to traverse hilly districts he proposed mounting

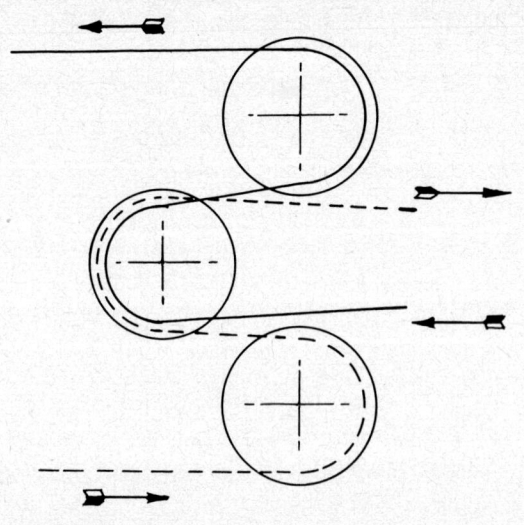

Fig. 85. Diagram of the rope drive system of Knight's digger

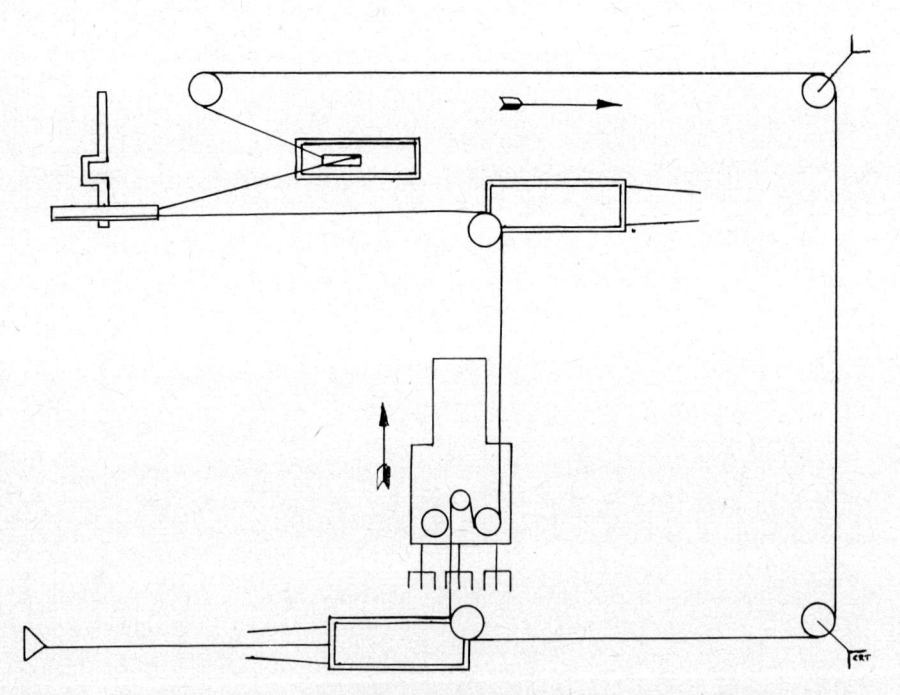

Fig. 86. Rope drive layout of Knight's system showing where the digger and rope carts were placed

two winding drums to the digger frame which, when driven by the flying rope, hauled itself along by one drum winding up its own rope and unwinding it from the other. A reversal of the drive hauled the digger back again. Secondly, he proposed driving the front wheel (now a single one as on Howard's machine) by means of a flat leather belt drive, which still enabled the digger to be steered by a tiller. Both modifications were superfluous as in the event the digging machine was not used on any but reasonably flat land where adequate grip could be obtained by the addition of "spuds" to the driving wheels.

Knight went on to patent an extraordinarily large number of ideas, mainly for oil and petrol engines. Some twenty of these were taken out by 1900, but no more patents for diggers appeared.

As we have seen, Knight's faith in his machine diminished and did not reach fulfilment in common with other exponents of tilling the land in an unaccepted manner, ending with the same financial loss and discouragement experienced by so many pioneers of steam cultivation. This might be understandable if the idea is a bad one, but if the idea is a good one—what then? Take the classic example of John Fowler, who had the expertise to implement and develop other ideas to perfection and further, to have resources left over to manufacture and market the product. It makes interesting speculation to wonder how Fowler would have treated the idea of digging machines had he lived long enough.

Knight went on to explore other areas of engineering and in 1868 made a small open four wheeled steam carriage, later constructing internal combustion engined cars. He died in 1917. As far as is known none of his digging machines survive.

CHAPTER EIGHT

The Early Rotarians

NO DOUBT THE READER would agree by now that the realisation of steam cultivation produced many strange looking machines of which perhaps the oddest were the diggers. Achieving a notoriety well in excess of the small numbers constructed, they can only be described as machines which caught the imagination of press and public alike, who, in seeking the spectacular were rewarded with being able to write and read about this wonder of the age. Who ever heard of a machine *walking*? When ever did an engine have *two* chimneys and work sideways on? How did forks and mattocks imitate the human action by means of gears and cranks? No wonder this breed of mechanical peculiarities received more than their fair share of publicity.

As far back as 1813 Richard Trevethick had designed a machine which he called his Steam Spade Tormentor (Fig. 87). It probably did not reach the constructional stage, but was a convincing design which may well have worked given enough power.

His famous experiments with high pressure steam are chronicled elsewhere, in any case being outside the scope of this volume. This mechanical genius could have been responsible for anticipating many of the ideas later taken up by various steam agriculturalists. His association with steam ploughing may well have been stronger than hitherto realized, only succumbing to the pressure of more urgent and easily realisable engines for use in potentially more lucrative directions. Not for the last time, simple lack of money prevented further development of steam cultivation, to be taken up more than twenty years later by John Heathcoat and others.

Trevethick's Tormentor consisted of a heavy wooden frame supported by four wooden wheels, and was equipped at the rear end with a huge wheel set at 90° to the frame. It appears that this fifth wheel was intended to be of cast iron construction with four tapered cruciform spokes. Attached to the cultivating wheel's rim were twelve scoops or spades and it was driven through a system of two each of bevel and spur gears, power to the shafts being provided by direct drive from the axle of the larger, rear wheels. Immediately in front of the spade or "tormenting" wheel was set a scoop of the same radius as the wheel. The presence of this scoop confirms that Trevethick was anticipating actual removal of the soil to one side leaving a shallow groove in the soil to be filled by the next pass of the machine, thus achieving inversion of the soil.

Power to the Tormentor would probably have been provided by an engine

Fig. 87. Richard Trevethick's steam spade "tormentor" of 1813. Although probably not constructed it was at least an idea for harnessing the power of steam and provided food for thought for other inventors
(*M.E.R.L.*)

which Trevethick seems to have constructed. Although details have not survived, the design of the Tormentor strongly suggests that it was intended to be drawn by rope or wire, making this set of steam tackle the first that could be so termed.

The engine would have been based on Trevethick's design of his rotative engine which was suitable for agriculture and other purposes (Fig. 88). Termed a "portable" it was not so in the sense that the word is understood now. Instead it was a stationary engine independent of brick superstructure or housing supporting any of the working parts. Thus the engine was free standing and only portable in the sense that it was of a size that could be loaded into a cart for transport from job to job.

Early in 1812 he built the first of these engines for Sir Christopher Hawkins Bart., M.P. of Trewithin, to replace the cattle mill previously used for thrashing. It was found that two bushels of coal costing 2/6d. ($12\frac{1}{2}$p) provided the same amount of work as four horses costing 20/- (£1·00).

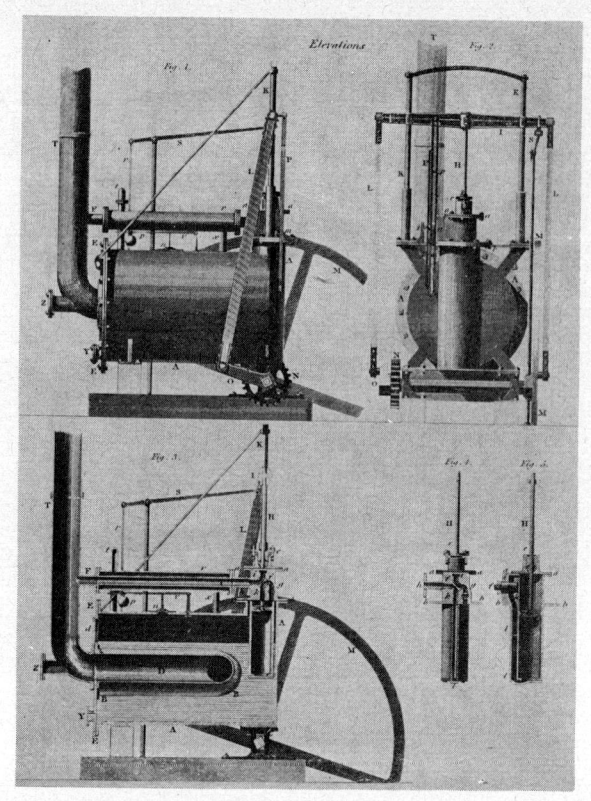

Fig. 88. Trevethick's rotative "portable" engine of c. 1816, of the type which would have powered the "tormentor", although this particular example was intended to power a dredging machine on the River Thames
(*Dr. Rees's New Cyclopaedia 1818*)

The success of this engine resulted in a second being made for Lord Dedunstanville of Tehidy Park.

Both of these engines had remarkably long working lives, the first working until 1879 and both being preserved in the London Science Museum, although only the boiler survives of the second. Further engines were made for Mr. Kendal of Padstow and Mr. Jasper of Bridgnorth.

Engines of this type were advertised as being suitable for thrashing and grinding corn, sawing wood, etc., weighing 15 cwt. at a cost of £63.

That Trevethick may be able to lay claim to making the first engine for exclusive use on the farm and even more interestingly, for steam cultivation is shown in a letter to Rastrick written in 1813—the same year as the watermark of the paper on which the steam tormentor was drawn—when he says:—

"I wish you to finish that engine with boiler, (i.e. the one to be sent to Exeter) wheels and everything for ploughing and threshing, as shown in the drawing, unless you can improve on it."

Now we have an engine complete with wheels—a true portable in fact.

A further step towards modern times had been taken by Trevethick in the following extract from a letter written in 1812 when he wrote to Sir Christopher Hawkins saying:—

"I am now building a portable steam whim on the same plan (i.e. as that of the thrashing engine) to *go itself* from shaft to shaft; the whole weight will be about 30 cwt. and the power equal to twenty six horses in twenty four hours."

Whether the two engines are one and the same is unknown, neither is it certain that they (or it) was completed, for no records have survived of the success or failure of the ideas, but certainly the letters suggest that construction was at least commenced on the first portable and self motivating engine. Incidentally neither idea being new to Trevethick as he had made a steam vehicle to run on the road with some success in 1801.

In another letter of 26th April 1812 to Sir John Sinclair, he says:—

"It is my opinion that every part of agriculture might be performed by steam; carrying manure for the land, ploughing, harrowing, sowing, reaping, thrashing and grinding; and all by the same machine, however large the estate. Even extensive commons might be tilled and effectually managed by a very few labourers, without the use of cattle. Two men would be sufficient to manage an engine capable of performing the work of 100 horses every twenty four hours, requiring no extensive buildings or preparations for labourers or cattle, and having such immense power in one machine as could perform every part in its proper season, without trusting to labourers. I think a machine that would be equal to the power of 100 horses would cost about £500.

"My labour in invention I would readily give to the public, if by a subscription such a machine could be accomplished, and be made useful."

Trevethick's offer of free services produced a resounding silence from the apathetic community, who were not inclined to offer money on a project which in their opinion was doomed to failure. As Trevethick was not rich, he required all his income from his mining interests for the maintenance of his family and therefore the first practical project as the development of steam in agriculture was heard of no more, an idea still-born for lack of the foresight with which Trevethick was endowed.

Perhaps the only other part of Trevethick's life relevant to the development of steam ploughing in particular and inventing in general, was the circumstances of his death on 22nd April 1833. He died a poor man, after a brilliant career which included many successful inventions, the building of many machines, development of the railways and locomotives, mining and many other projects, yet he was borne to a pauper's grave by mechanics from the works of Halls of Dartford. Buried in an unknown spot allotted to the poor, the turf was replaced without a stone to mark the great man's last resting place. Why was there no stone? Because subscriptions raised to furnish one for the grave did not amount to enough to buy one! Strange how this pattern is repeated, to lesser degrees throughout the age of invention. Society was indeed ungrateful to its members who strived to improve its lot. Perhaps they were just not suited to the world of business, preferring to explore the intricacies of mechanics, in preference to making their fame and fortune.

Although something of a combination between a rotary and digging action Trevethick's Tormentor was the first practical thinking about departing from the traditional plough—others were to follow.

Digging exponents diverged into two separate streams of development, one favouring a mechanical reproduction of the motion of the spade as manipulated by human hands, the other seeking to find a means of digging by other methods which were usually rotary.

Between 1810 and 1897 there were a number of rotary cultivating engines proposed which are historically classifiable under the general heading of minor attempts to cultivate the soil by steam power using other means than the plough or fork. It may well be that that it is something of an injustice to state this, as limited knowledge of their various activities may leave an impression of obscurity and failure while the better publicised attempts leave us with more information to record, in spite of the fact that they may have been less successful technically speaking.

The early ideas of Trevethick were preceded by Major Pratt in 1810 when he stated:—

"And also, by this my machinery and mechanical powers, I do away with the necessity of keeping animals and prevent thereby the consumption of agricultural produce, and cause agriculture to operate in various ways to the good of mankind in more amply supplying the markets (Fig. 89)."

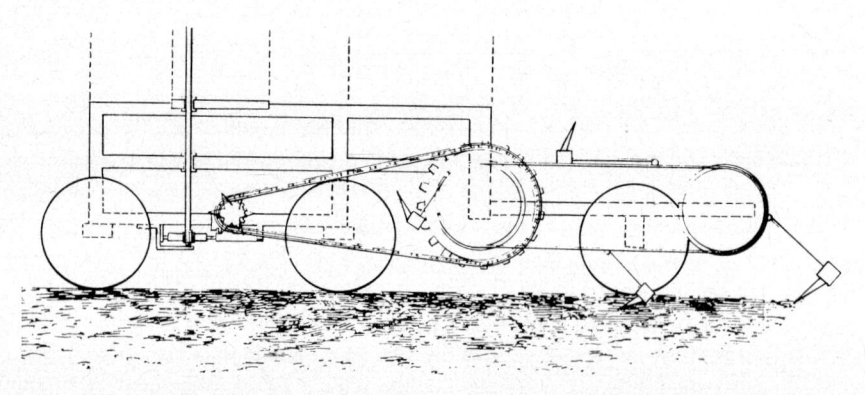

Fig. 89. Pratt's mattock equipped digger of 1810 is almost identical in principle to R. A. Clarke's digger (which see) except that the belt carrying the mattocks is at the rear of the engine, instead of directly underneath
(*Transactions of the Society of Engineers 1868*)

So, hopefully does Major Pratt express his ambitions and the ambitions of decades of inventors to come for the mechanising of agriculture in his Patent specification of 1810 for his cultivating machine.

He was probably the first to set down on paper ideas to dig by means of a power other than the manual labourer, his machine being an interesting combination of rotary cultivator, digging machine and cable ploughing apparatus embodying most of the ideas later used by other inventors of steam cultivating tackle, in one way or another.

The Patent specification is a somewhat jumbled mass of drawings, expressing basic ideas clearly enough, then frustratingly leaving out details which would usually have been considered essential to complete the picture. For example having described the various implements he says "I have chiefly given the ploughs for example, but harrows, rollers, hoes, drills, or any other implement of husbandry, placed on or fixed to this or any of the machines, will perform their operation in like manner, being connected with wind, steam or any such like mechanical power, which any ingenious workman will know how to do."

Thus does he leave any implement to be powered by any power leaving one wondering whether the Patent would not have been better taken out by the "ingenious workman"! In fairness it must be said that in 1810 the high pressure steam engine was in embryonic form and it was to be some fifteen or twenty years before compact, reasonably reliable engines appeared.

His designs were based on wooden frames to which were attached the implement which in turn were moved by alternative means.

The first method of motivation was by means of chains between two carriages hauling further carriages to and fro—the system later commonly known as cable ploughing. The second method enables Pratt's ideas to be included in these pages as a digger. A wooden carriage was provided with two horizontal wheels. Upon each wheel was mounted a device now termed a swash plate which carried two small plough shares. As the wheel revolved, each share was plunged into the soil and out again by the inclined angle of the swash plate. As the carriage advanced, the two wheels, revolving in opposite directions, left a cultivated swathe which must have had the appearance of the surface of a piece of metal milled with an end cutter. The carriage was moved forward on its free running wheels by a series of mattock shaped claws on a chain which dug into the ground and hauled the device along.

Pratts designs were not constructed, remaining paper dreams, awaiting the advent of the long awaited power of steam. The reign of George III was not to see the breakthrough of mechanical cultivation and the reigns of George IV and William IV were to pass until the beginning of Queen Victorias reign, before John Heathcoat made his practical attempt at steam ploughing.

Twelve years after Major Pratt's proposals, M. J. Roberts designed a digging machine which was probably horse powered, but was significant in the history of digging machines in that it was the first attempt to mechanically reproduce the human action of using a spade. Few details are known of this enterprise.

A rotary digger in model form, was made by T. Atkins of Chepstow in 1843, and exhibited at the Shrewsbury meeting of the R.A.S.E. Produced according to the patent of an American, J. A. Atzlar, the principle of the drive of this system was not unlike Knight's hop digger, made some 36 years later. Specifically for the purpose of cultivating hilly terrain, the rotary digger was an independently moving carriage operated by a rope, which after passing round a grooved pulley on the portable engine went to the pulley wheels of a "rope carrier". The rope carrier possessed a "weighted pulley, rising and falling according to the fluctuations in the tension of the rope", which ensured that a steady strain was imparted to a travelling carriage to which was attached the rotary digger, which could proceed over uneven or flat ground with equal ease.

The rotary digging drum was of unusual construction in that additional smaller drums mounted around the main drum carried the "spades, knives, prongs or teeth, as required". Other operations which the carriage was claimed capable of performing included "cultivating, sowing, reaping, mowing, dressing, ploughing or other work required in agricultural business".

In a lecture to the Society of Arts in February 1856, Atkins said that the distance between the carriage and engine could be as great as 1500 yds., claiming that the rotary cultivator could work up and down the land and that 1,000 yds. of land could be cultivated with 10 cwt. of rope and a 20 or 25 H.P. engine.

A full size machine was made, and worked at Blackthorn, near Bicester, Oxfordshire. Provisionally patented in August 1853 and February 1854, Atkins described the digger as having rotating forks or spikes, presumably these were the type actually used on his digger, which apparently was quite successful in breaking up and loosening the soil. He considered that the best means of propelling the digger would be "to have a large fixed engine of 20 or even 30 horsepower, set down in the centre of a farm, driving, by means of endless wire ropes, extending to a great distance, and working at a high speed, so as to diminish the weight of rope required, the rope being carried on standards at some height from the ground, like a colliery rope. This idea had been earlier suggested to him by seeing a rope manufactory at Bristol, where a rope of two miles length had been working constantly for two years, taking the power from a 10-horse engine, and driving various machines all down the walk for a mile distance."

In a lecture to the London Farmers Club in 1862 James Howard in attempting to probe into the reasons for the failure and of apathy towards so many ideas of different types of steam cultivation machines said: "Political economists tell us that the machinery of a country will naturally correspond with its wants, and with the history and state of its people. This is undoubtedly true, the schemes we have described having been invented before they were really wanted or before their need was felt. For instance, not longer ago than the Shrewsbury Meeting in 1845, a model of Atzlars American steam plough was exhibited in a room at Shrewsbury, and the town placarded informing the visitors of the fact, and yet no one went near except two Russians who dropped in towards night to look at it."

So Atkins' model and equipment were another to suffer the fate of many of his contemporaries. Lack of interest ensured that no more was heard of either Atzlar or Atkins.

One of the "in between" types of machine, being neither a true digging or

EARLY ROTARIANS

ploughing engine was introduced by Sir John Scott Lillie in 1846 (Fig. 90).

Bearing a resemblance to Heathcoats caterpillar tracked engine of 1837, Lillies version was similarly tracked, but incorporated a set of four wheels within the framework set at right angles to the direction of travel. These wheels could be jacked down to below the level of the skeleton drums or wheels carrying the caterpillar tracks, lifting the engine bodily up and enabling the whole machine to be moved sideways on to uncultivated land. The operation was then reversed, the engine set down again, proceeding on its way cultivating the next strip. A neat way to overcome turning at the headlands.

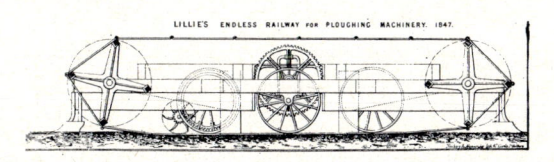

Fig. 90. Sir John Scott Lillie's massive machine of 1847 was a combined ploughing and digging engine. The rotary cultivator can be seen to the right of the left hand main wheel
(*Transactions of the Society of Engineers 1868*)

Cultivation was by means of a rotary cultivator approximately amidships and set between the tracks. It was driven by internally toothed gears, inside wheels which assisted in supporting the rather flimsy looking tracks. The principle is best demonstrated in the working of a modern side wheel, hand pushed lawn mower, where the cutter is replaced by the cultivator. Coulters or cutting irons attached to the frame of the engine preceded the cultivators assisting in initially breaking up the soil. A double mould or ridge plough was attached to each end of the carriage framework, clearing the way for the main wheels of the engine.

The power unit—which would have been very useful to Major Pratt and other early designers—was a "high-pressure steam engine on the carriage frame for the purpose of impelling the machine, by turning a whelp wheel, and causing it to draw in a rope attached to grapnels in each side of the field". So the engine was not entirely self moving, indeed this stage of development had been reached by few engines to this date. It is possible that it would have been an unnecessary luxury on Lillies engine, as it was probably used entirely on the same area, not requiring to be moved from job to job, as with later engines which had to be more mobile.

Among the few representatives of steam digging outside the United Kingdom were the French brothers Barrat (Fig. 91). In 1847 Pierre Philippe Celestine Barrat patented an engine which operated in novel fashion.

Popularly termed a steam grubber, the machine mechanised the operation not of the fork or spade, but the ancient implement allied to the adze—the mattock. One of the first completely self moving engines, it was constructed in Paris, and was nicknamed "la piocheuse" ("the pick axer"). The design was "in an arrangement of one or two rows of teeth, to which, on the one hand, a

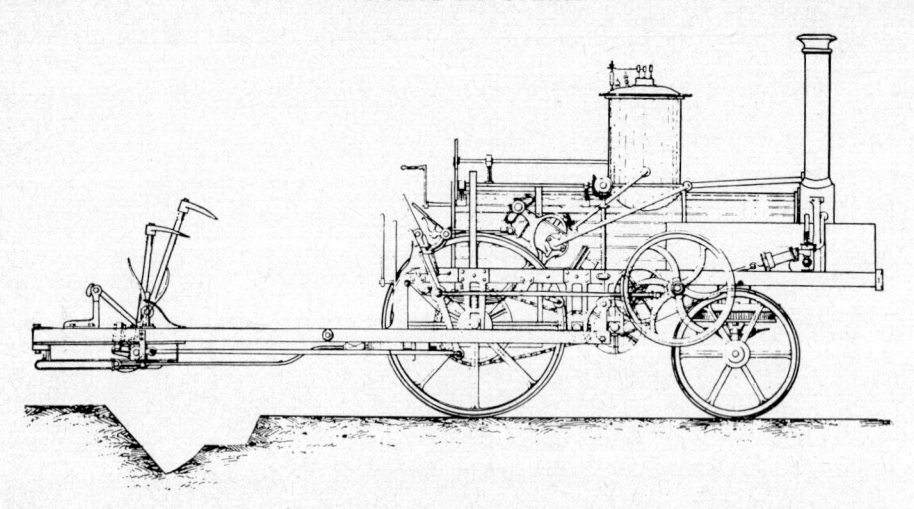

Fig. 91. One of the few diggers to employ mattocks, Barrat's engine of 1847 was self moving, while the double row of mattocks hacked the soil with a reciprocating motion to two depths
(*Transactions of the Society of Engineers 1868*)

rotary alternate movement is given and on the other hand a to and fro movement, in such a manner as to imitate as much as possible the work of the hand with a mattock, whereby the earth is returned to the ground after it is raised".

A Company said to have a 1,000,000 franc capital was formed to promote sales. The number of "pick axers" made is unknown, but enough were distributed throughout France to make it a well known engine.

The boiler was wood lagged and mounted upon a frame to which the two large rear and two small front wheels were attached. The engine was an undertype with the cylinder mounted in the frame under the boiler.

An extension of the frame or chassis to the rear carried the mattocks, arranged in two sets of different lengths mounted alternately on the same shafts within the frame. As they hacked the soil "The mattocks of the foremost row enter the earth to the depth of from 7 to 12 in., according to the nature of the soil; while those of the back row enter in the place where the front mattocks had previously operated, and penetrate the earth to a greater depth". This depth of penetration was undoubtedly an advantage over other forms of digging or ploughing and must have been of great benefit in breaking up the subsoil. Each mattock had a fished head mounted on a straight wooden shaft which in turn was inserted into a metal socket on the horizontal shaft, about which it freely pivoted. Each mattock had independent movement and to complicate the design, the shaft carrying them also had a longitudinal movement in guides attached to the digging-frame—in the manner of a sliding seat in a modern racing shell. All this in addition to the "alternate circular motion"! Penetration of the mattocks was adjusted by means of altering the pressure of springs acting upon them (Fig. 92).

Fig. 92. Barrat-freres digging engine. This engine apparently combined reciprocating mattocks with rotating cutters

Other foreign makers of digging machines, included Broomen, Jeannevet, Dussuc and Bauer but the writer has been unable to discover any details of these shadowy figures.

A novel apparatus was designed and patented by George Calloway and Robert Purkis in 1849. It is almost certain that this machine did not progress beyond the model stage of which more than one were made in 1851.

The twin cylinder vertical boilered engine was mounted on four wheels, a manstand at the front. Driven by gearing from the rear wheels the cultivating mechanism was mounted across the rear of the engine. This consisted of a flexible belt turning about two wheels set some 18 ft. apart on an iron frame in caterpillar track style. The belt was made up of rods connected by links similar to a chain. Equally spaced around the periphery of the chain were four small shares and coulters mounted rigidly to it. As the chain revolved so the shares cut four furrows for each complete revolution, the engine in the meanwhile advancing 3 ft. This resulted in the furrows being at an angle to the direction of travel. The cultivating chain could be raised out of and lowered into work by means of the rotating screwed columns which were hand operated by crank handles. To decrease the width of the engine for road use the iron frame carrying the cultivating chain was hinged either side and by means of two more cranked handles could be raised into vertical positions.

A Company rejoicing in the name of Keddy's Patent Traction Engine, Steam Plow and Land Cultivation Co. advertised Thomas Keddy's rotary cultivator in 1859. From Handsworth, Staffordshire, Keddy patented his

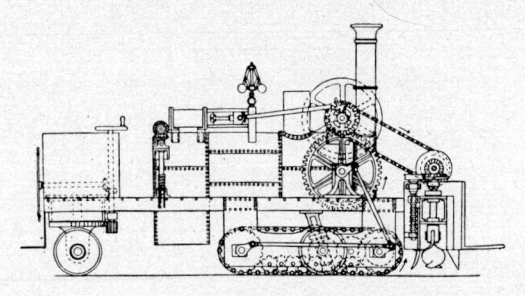

Fig. 93. Keddy's American built steam cultivator, in principle not unlike the Fowler Gyrotiller
(*Abridgements of Specifications*)

somewhat ugly engine in 1857, advertising it for hire or sale quite extensively in America. The Steam Plow Co. was formed with a capital of $50,000 and aimed at promoting his ideas.

Vertically mounted, contra rotating screw tillers mounted behind the caterpillar tracked engine performed the digging, giving the machine the appearance of a steam powered Gyrotiller (Fig. 93).

As Keddy's digger was advertised for hire to dig by the acre in Illinois, presumably at least one was constructed.

CHAPTER NINE

The Later Rotarians

OF THE MANY INVENTORS to propose various methods of steam digging a comparatively small number actually made, tested and used rotary digging engines.

Some interesting designs resulted with varying degrees of success and failure in their application.

The following pages describe the exploits of the known exponents of the art between 1849 and 1856. Strangely, after this date there are no records of rotary digging engines being constructed. It was not for some twenty years that further attempts to dig by steam were made and then it was by means of reciprocating, forked machines.

James Usher

A Scottish representative of steam digging, James Usher of Edinburgh introduced his rotary machine in 1850 in the form of a model which can be seen today exhibited in the London Science Museum (Fig. 94). First shown to the Highland and Agricultural Society the model was also on show in 1851 at the Crystal Palace. Fully working, Usher's model engine is spirit fired, made largely of brass, and painted bright blue on the boiler, black smokebox and firebox with an elaborate red floral design worked on. The words "Patent Steam Plough. James Usher, Edinburgh. Patentee" are emblazoned on the boiler barrel in gold, completing an extremely attractive miniature rotary cultivating machine, alone worth a visit to the Science Museum.

By the end of 1851 a full size engine made by Slight, also of Edinburgh appeared. Weighing $6\frac{1}{4}$ tons, of 10 H.P., the engine was not dissimilar to a railway locomotive of the period. Unlike most rotary digging engines, the cultivators were mounted at the smokebox end, a modification made between building the model—on which the cultivators were attached to the firebox end of the engine—and the full size machine. Travelling backwards, the digger left a swathe 4 ft. 2 in. wide (Fig. 95).

At the International Exhibition of 1862 "direct acting rotary engines" were represented in the Agricultural and Horticultural Machines and Implements section. Unfortunately, these are not detailed in the record of the exhibition, but in the historical notes appended on steam ploughing the following passage on Usher's machine appears: "Mr. Usher of Edinburgh devised a

Fig. 94. James Usher's rotary cultivator in model form
(*Crown copyright, Science Museum, London*)

'locomotive engine', to the hind part of which a moveable skeleton drum was attached, carrying on its outer circumference a set of four or five plough breasts, each covering a space of ten inches, and being arranged in a spiral form around the cylinder, so that each plough followed behind the other, taking a different cut of the same width, and effecting the ploughing or reversing of the surface in a diagonal line. This skeleton drum was driven round at a slow speed from the spur wheel of the engine, and by working against the soil at the rear propelled the machine along in a straight line. The engine was a light tubular form with a pair of fore carrying and turning wheels, but constructed with a broad iron roller instead of wheels behind, by which increased friction was obtained while the pressure per square inch of the engine on the soil, by being distributed over a greater surface, was materially lessened".

Self moving gear was included, but almost certainly used merely to move the engine from job to job.

The first engine was demonstrated at Inverleith and met with considerable success, cultivating at a speed nearly twice that of horses at an estimated cost of 3s. (15p) per acre, compared with 10s. (50p) by horses. Further demonstrations were given in the spring of 1852.

The steersman walked alongside the engine, turning a small crank one way or the other, revolving a large turntable the width of the engine, upon which

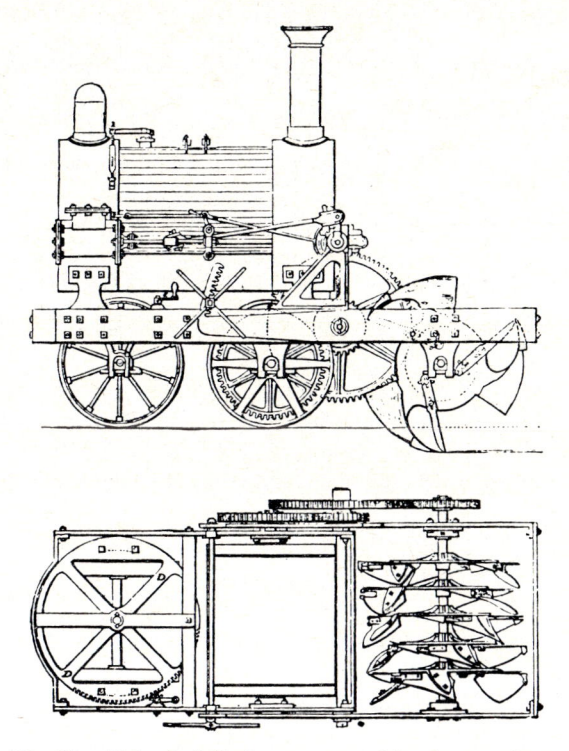

Fig. 95. Usher's full size rotary cultivating engine

were mounted two narrow tracked wheels, an arrangement convenient for turning the engine almost in its own length, but a somewhat tedious process for the steersman.

Two cylinders placed one each side of the firebox turned the driving cranks through very long connecting rods. Interestingly, no flywheel is apparent on either the model or contemporary drawings, which must have resulted in extremely jerky motion. A capstan operating a rack and pinion raised and lowered the cultivators. Before commencing work, in the absence of a fly wheel, the cultivators must have been allowed to gather speed before being lowered gently into work.

Another engine probably of the same type was made in 1853 and yet another in 1854 but this time modified with a vertical boiler with greater heating surface. Weight was reduced by one ton and the cultivator could be lifted in and out of work by powered means, probably hydraulic. It could cut to a depth of $9\frac{1}{2}$ in. at 30 r.p.m.

Writing to J. A. Williams—a fellow pioneer of steam cultivation—James Usher says of his latest engine: "The drawbacks I had in my first machine were want of steam power and too much weight. The first of these I have obviated in the new one by having 210 feet of heating surface in the boiler, against 130 feet in the last one, which gives six horses' more steam; and the latter objection

I think will be entirely overcome by this additional power. The work done I think is nearly perfect . . . and the expense per acre is about one third of the cost by horses."

J. J. Mechi as with Romaines machine—and with customary enthusiasm—took an interest in Usher's venture commenting in 1855 at a meeting of the London Farmers Club: "He had seen the model of Mr. Ushers steam cultivator and he had a strong opinion that, on account of its slow motion, it would effect the object which Prof. Wilson had so well described—the perfect cultivation of the soil. Cultivated land was not like a common road. An engine on common road was designed to move at a rapid rate, and its tendency was of course to destroy the road—an effect it was desirable to prevent as far as possible. But in agricultural operations the very thing they wanted to do was to destroy the land, and break up the soil; and the very act of getting a leverage on the soil was a means of propulsion, provided only there was the requisite power in the engine. He was not in a position to state which would be best; but it struck him, if they could get the engine to go with the plough, that that would be by far the best way."

At this point a member of the club audience shouted "You will never do that" and J. A. Williams agreed saying "Never".

Unperturbed, Mechi continued "It had been done in the case of Mr. Usher, for he had not only ploughed his land, but his engine had come back over the same land, and ploughed it again. The experiment had been well recorded. . . . Mr. Usher was, he understood to appear at Carlisle next July, and he wished him success, for that gentleman had spent a good deal of money upon his experiments; and if it were necessary, he (Mr. Mechi) should be glad to contribute towards forwarding the model to the show".

In the event, Usher did take his engine to the show at Carlisle.

Trouble started at the show ground when the engine promptly bogged down. When eventually extricated, it was driven to the trial ground where it was found that the plot of soil to be cultivated was less suitable for working than soils previously tilled.

Criticism inevitably followed. "The machine under the most favourable circumstances, could with difficulty move itself, and the revolving shares neither inverted nor pulverised the soil but tumbled it about in wild confusion, and left it in a state more unfavourable for cultivation than it was before."

Ushers intention had been to cultivate the area allotted to him, finishing off around the edges of the field, but—indicating a lack of power—the engine was unable to cope with the sloping ground.

While having some justification, the critics remarks were in some ways irrelevant for it had never been Ushers intention to work land of this type, but land which had already been cultivated and reasonably flat. He readily agreed that his engine was not capable of tilling all types of soil and that at £400 the engines were expensive. He considered that the savings over the expense of keeping horses would be considerable, also that an even greater saving would be made by adapting his engine for other purposes such as sowing, harrowing, threshing and even possibly reaping, but as far as is known, no such adaptations were made by Usher.

Regarding the capital outlay he said "I think if the farmer cannot afford to lay out so much money, the landlord will readily do it, and then gentlemen, we may defy the world for producing cheap corn".

Neither farmer nor landlord were convinced sufficiently to invest in Ushers engine, although he was confident enough to form a company for the purpose of promoting his invention, but by 1859 both the company and the engines had submerged beneath the numbers of succeeding ideas, all claiming to be the salvation of agriculture; a few of them being rotary digging machines. A valiant little group, Usher ranks among them as one who made a brave attempt but in failing, he at least left us with the little brightly painted model of his engine at South Kensington.

Thomas Rickett

Among the more notable if not the most successful pioneer—in so far as any digging engine pioneer can be called successful—of rotary digging engines was Thomas Rickett of The Castle Foundry, Buckingham. He was one of the few who ventured into developing the steam engine, not only for the purposes of cultivation, but also for the application of his designs to passenger carrying road vehicles. It is for his steam carriages that he is better known today.

His work was of advanced design (Fig. 96), the rotary digging engine was of handsome appearance, generally following the lines of contemporary makes of

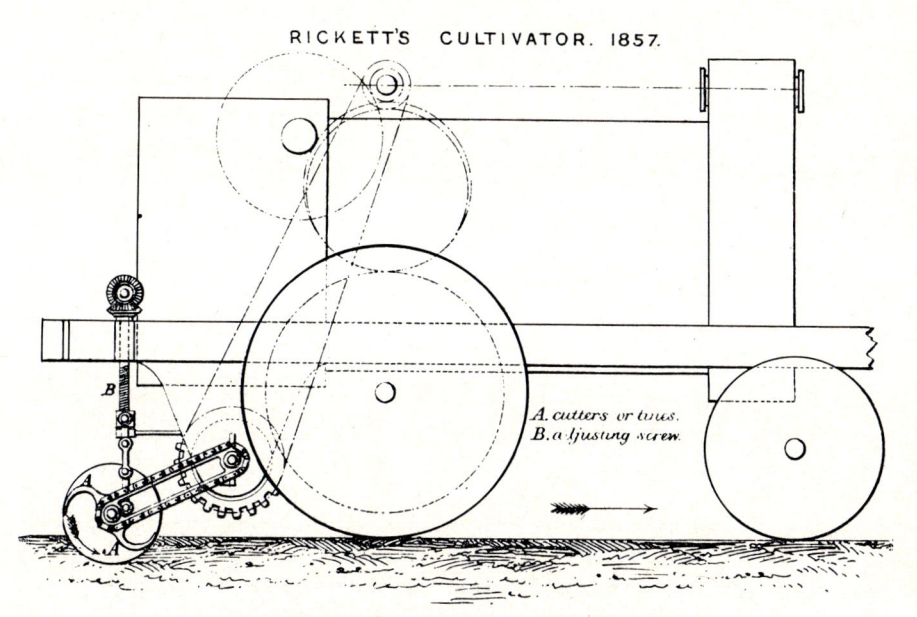

Fig. 96. Thomas Rickett's cultivator
(*Transactions of the Society of Engineers 1868*)

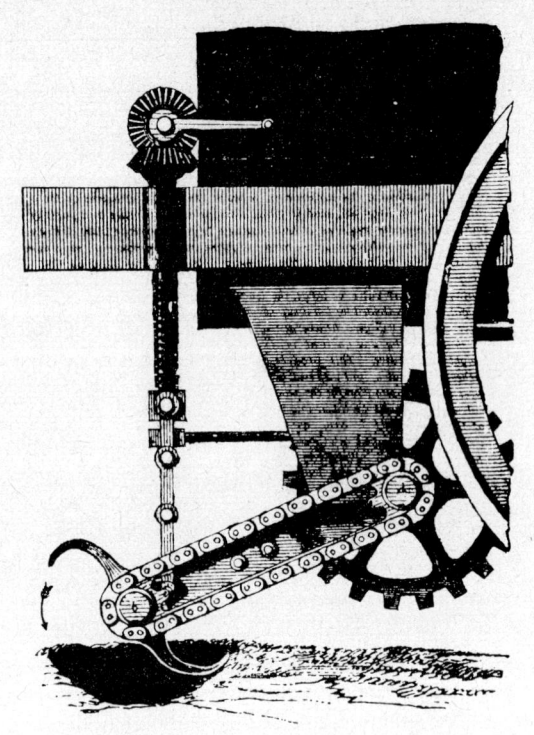

Fig. 97. The chain driven cultivating mechanism of Rickett's engine of 1857. It was this chain which broke during trials

traction engines. Following the practice of Aveling and Porter, the twin cylinders were built in to the smokebox, the crankshaft driving a chain of gears which in turn rotated the cultivator through a chain. The rear wheels were 4 ft. diameter by 12 in. wide, and due to only one being driven, some adhesion problems were experienced, although Rickett claimed that the engine had worked on a stubble field with a gradient of 1 in 7.

The cultivator (Fig. 97) was situated at the rear of the engine and in this respect was perhaps improved or at least different from other rotary digging engines in that it rotated in the opposite direction to the travel of the engine, thus effectively biting the soil from the deepest part of the trough made by the cultivator, and actually lifting it, carrying it right over and depositing it behind the engine. Advantages claimed were "that the cutters are better able to cope with strong or very hard soils and on entering the unmoved ground at full depth, they avoid the difficulty of penetrating from the top when dry and baked in the sun".

While overcoming the objections of some—that rotary diggers did not invert the soil—it was quickly pointed out by critics—of whom as always there were many—that the subsoil which the engine had just passed over was still untouched, possibly being even more compacted than before.

The first engine weighed about 6 tons and was described as being of "rough style", no doubt bearing all the marks of experimentation. It possessed only one forward speed and unlike Usher's machine, was driven forward while cultivating behind. Change wheels were provided to allow different speeds of cutting the soil. These were situated between the crank and intermediate shafts.

Some alteration of cultivator height was provided by means of a crank handle working through bevel gears. Again unlike Usher's machine there does not seem to have been any self adjustment of depth of cut, which was to prove a disadvantage in trials.

Steerage was from a platform over the front wheels by means of a hand crank. The platform also served for stoking, but where the coal was carried is a mystery, presumably enough for a few shovels being kept on the platform floor.

Ricketts engine was shown in 1858 at the Chester meeting of the Royal Agricultural Society of England, with great hopes of success (Figs. 98 & 99). These were speedily dashed when the chain driving the transverse shaft on which the two cutters were attached, broke early on in the trials, resulting in a disappointed Rickett being compelled to withdraw from the competition.

"The field appointed for the trials was laid up in the narrow lands, varying from 7 to 10 ft. wide, with a difference of level of some 10 or 12 in. between the ridge and the furrow, and thus afforded no fair test of the powers of this rotary cultivator as, owing to the great breadth of its cut (7 ft.), in no part could it find space sufficiently level to show its work with advantage."

However, enough work was done to favourably impress the judges with the possibilities of the engine and they appraised its capabilities and said—no doubt with a glance at the broken chain—that several points in its working detail could be improved, even adding a careful statement to the effect that the principle of rotary cultivation "had taken a distinct position as a desirable and valuable addition to the Mechanics of Agriculture", commending Rickett for his skill and ingenuity of design of the rough trial machine.

Details of Ricketts first engine were: While moving on suitable ground, the engine advanced at the rate of 20 ft. per minute, the cultivators working at 75 r.p.m. taking a slice $4\frac{1}{2}$ in. wide, 6 in. deep and 7 ft. broad. This gave a rate of $91\frac{1}{2}$ poles per hour or about $5\frac{3}{4}$ acres per day.

Expenses were calculated thus:—

	s.
Engineer	5
Two men at 3s. (15p)	6
Coals	10
Oil etc.	1
Water cart	5
Interest per cent., and wear and tear	8/9
15 per cent on first cost, taking 200 as the number of working days per year.	
	£1 15s 9d (£1·79p)

Fig. 98. Rickett's handsome cultivator of 1857. Note the twin cylinders forming part of the smokebox and forecarriage steering, both common characteristics of traction engines of the period
(M.E.R.L.)

Fig. 99. Perhaps elegance was concentrated on too much in design, resulting in some parts being too lightly made to withstand the ardours of digging
(M.E.R.L.)

Taking the average area cultivated per day as 4 acres this would give about 9s. (45p) per acre cost. Rickett was encouraged enough to start work on a modified engine, in an attempt to eliminate the faults of the first one.

He wrote to J. A. Clarke a letter which was included in Clarke's prize essay "On the application of steam power to the cultivation of the land saying:

"I can hardly consider my present arrangements improvements upon the one you saw at Chester, as in that case the cultivator was adopted to an engine as an experiment, the successful issue of which led me in the next place to adapt the engine to the cultivator or to design a machine specially for locomotive steam cultivation. Such an engine I am now building: it is 15 H.P. and will cultivate at least one acre per hour 10 inches deep. The cultivator shaft is driven by coupling rods direct from the engine crankshaft, both shafts making 60 revolutions per minute, and the engine having a 22 inch stroke. This is the minimum speed, a much greater velocity being practicable when not inverting. There is an arrangement of land wheels to maintain always a uniform depth of digging and only two men are required to work the machine. But the point to which I mainly directed my attention was that of providing an extended bearing surface on the land, believing that, although there are times when rolling wheels do not materially compress the land, generally speaking their effect is injurious. I sought in fact to bring the only real objection that can be raised to direct action.

"I patented last autumn a plan of an elastic roadway by which I distribute the weight of the engine over 4 or 5 square feet and therefore reduce the pressure on the land to about that of a mans foot. The price of this engine will be about £600. I propose using 18 inch tines in this cultivator, the diameter of the circle they describe being three feet: and two cuts in each revolution of 10 inches each. The width is 7 ft. 6 in."

This new engine was exhibited at the Warwick show of the R.A.S.E. in 1859 as a static item, leaving us with no information as regards its performance. It had twin cylinders 8 in. by 22 in. and was an undertype.

The patented "elastic roadway" was made in three sections joined together by cross pieces which acted as strakes, and preventing the tyre from damage. The strake ends were formed inwards to prevent ingress of dirt to some degree into the gears which meshed with the internal teeth of the tyre. The principle is similar to the model of Whatley's plough, made in the early 1930s, in the authors possession. Whatley's principle is very similar to the "elastic roadway" and may indeed be either a resurrection of Rickett's idea or have been inspired by it.

Rickett's new engine had main wheels of 7 ft. diameter which were driven by spur gears on each end of the crankshaft, which meshed with the internal teeth of the main wheels. Two other wheels running inside the tyre and positioned either side of the main wheel ensured that a flat section of the tyre was in contact with the ground. Ricketts claim that the weight of the engine was spread over some 6 square feet would on further investigation seem to be a misnomer. The flexibility of the tyre, so essential to its working, made quite sure that no weight was taken on the flat part of the tyre except where the three wheels each side made contact with the ground through the tyre. The main benefit from the idea would be the facility of the engine to ride over obstacles in its path, but as most fields are fairly flat it would seem to have been an unnecessary luxury.

The cultivator was mounted as before at the rear, but now shaft instead of chain driven, and raised and lowered by steam or hydraulic cylinders. A wheel running on the land side of the engine, attached in line with the shaft of the cultivator, raised and lowered it, compensating for undulations in the ground, keeping a constant depth of cut. Width cultivated was 7 ft. 6 in. and depth claimed 10 in.

No more is heard of the fate of these two engines, but they could not have made the impression that Rickett hoped for. He went on to design and manufacture light steam carriages, which met with greater renown than his attempts at steam cultivation.

A "direct acting" engine, made in 1864 does not seem to have been intended in any way for cultivation, but for fast road haulage. The direct action was between the two 8 in. diameter cylinders and the rear wheels, achieved by eliminating all gearing in between. Demonstrated "after repeated trials of a great many different descriptions of road locomotives appears to have taken the railway locomotive as its base. . . ." Rickett claimed in a letter to *The Engineer* that his engine was capable of ascending gradients of 1 in 10 or 12 and guaranteed that it could draw four tons at 10 m.p.h. although capable of greater loads under the right conditions. Little more is heard of this enterprise and Rickett did not build any more digging or traction engines.

Why he was described as Manager of Castle Foundry is something of a mystery in itself (Fig. 100). Figures that one usually associates with such a concern are in the shadowy background.

It is thought that a Mr. Beards of Stowe, for whom he made a steam carriage, was connected in some way with the Foundry other than just a customer, but again details are unclear. Whether he had a financial holding in the concern is unknown, as is the eventual fate of Rickett. All patents connected with the Castle Foundry are in his name, suggesting that he may have had some sort of holding, and in todays terminology may have been designated the Managing Director. When and for what reason he left the Castle Foundry

Fig. 100. A steam carriage constructed by Thomas Rickett. He made at least two; one for the Marquis of Stafford in 1859 and one for the Earl of Caithness in 1860. The second carriage is shown here with the Earl driving

is not known, but on 23rd June 1865 a sale was held closing an era in another nineteenth century iron works. Rickett moved on to Islington in 1865 and Birimingham in 1869 from where a provisional patent was issued for a velocipede, a rather inefficient looking two wheeled carriage.

John Bethell

In 1852 a rotary digging engine patented by John Bethell of Westminster, London took the form of a modified portable engine (Fig. 101). The digging mechanism consisted of a rotary cultivator turning "in bearings made at the end of a frame, or a pair of arms secured to either the front or back part of the carriage in such a manner that it may be raised or lowered as may be required; there is also a set screw for securing the frame and digger at any given point. On one end of the digger, shaft or drum is mounted a rigger or pulley, to which motion is communicated by means of a band or strap passing from another rigger actuated by the engine. When the machine is required for digging or cultivating land or level ground, the rotary digger is lowered on to the ground, and upon rotary motion being communicated to it, the tines will cut away the ground and throw it backward.

"While the machine is in operation, it is drawn slowly over the land by horse or other power; and by that means the digger will be continually operat-

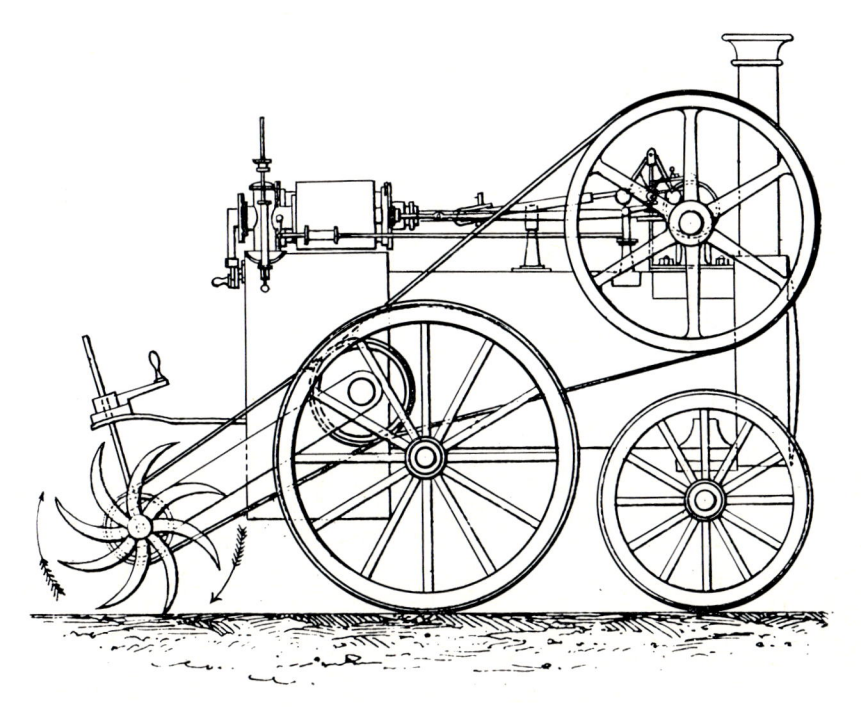

Fig. 101. Rotary cultivator designed by John Bethell in 1852.

ing on fresh ground, and the earth, when thrown backwards, will be deposited in a finely divided state, and in an even and regular manner."

So went the words of the patent, which describes the idea adequately.

From a statement made by Bethell at the Society of Arts in January 1856, it seems that a portable had been suitably modified and worked for three years, with various modifications during the period which from time to time suggested themselves.

It was stated that "it dug like Parkes steel fork, and left the ground in a perfect state of tilth after the separation. It threw the earth up into the air, the earth falling first, being the heaviest and weeds being left on the surface. They had no difficulty in working four to five acres in a day with that machine, digging down to a depth of nine inches." Witnessing farmers stated that it did as much in one operation as would require two or three ploughings to perform, as well as scarifying and harrowing. Cost of running the tackle was estimated at about 9s. (45p) per acre, whereas 23/– (£1·15) was stated to be the cost of animal power. The experimental engine worked at a pressure of 45 lb/sq. in.

An interesting secondary use of the digger proposed by Bethell was that it was used for "excavating or cutting down hills for the formation of railways . . ." when used for this purpose "the digger is placed in front, and is pushed forward up to the hill or bank, either by mechanical means or in other suitable manner, and the rotary cutter, being made to operate against the hill or bank, will cut down the earth and throw it under the carriage, and from thence it can be removed either by a chain of buckets or by hand labor, and deposited in trucks or carts for removal to a distance".

Whether this idea was put into practice is doubtful, but is an early indication of thoughts turning to other potential usages of steam power, this particular one although abortive, finally matured in another industry with the introduction of electric power in the mines.

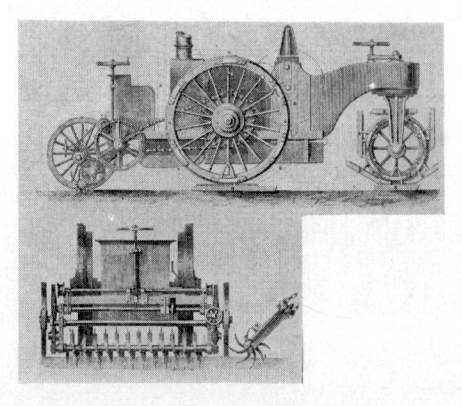

Fig. 102. Bethell's second and improved engine of 1857. Boydell wheels were employed to distribute the weight of the heavy machine
(*The Engineer 1858*)
(*M.E.R.L.*)

A rather elegant looking self moving engine was patented by Bethell in 1857 (Fig. 102). There was something of a graceful appearance about the swan necked fore carriage, although it is certain that it was that shape for practical purposes only, and indeed merely provided clearance for the Boydell "Elephant Footed" single front wheel. The two rear wheels were also equipped with the heavy shoes, approved of in Bethell's case by J. A. Clarke in his prize essay "and certainly the addition of endless rails or shoes of some kind, is indispensible for bearing up the weight of a ponderous locomotive engine and its machinery in order to prevent injury to the condition of the soil".

As before, the rotating digger was placed at the rear of the traction engine and was now chain driven initially from the rear axle, and finally by gear wheels. A refinement was the addition of a land wheel which rode over undulations in the ground, keeping a constant depth of cut.

There is no record of Bethell having made any other digging machines.

Robert Romaine

Robert Romaine from Peterborough, Canada, arrived in England during 1853 and commenced work on his rotary cultivating engine. He obtained a patent in which he stated that his cultivator could be used for pulverising, levelling, drilling and rolling the soil in addition to the prime purpose of digging. Other jobs proposed were "sowing, depositing manures, reaping and mowing, making drains, and performing other agricultural operations by mechanism actuated directly by a steam engine, but transversed or conveyed over the land by horse power or by the power of an additional steam engine". In other words a horse drawn cart containing its own engine for performing the various tasks described (Fig. 103). In the event, it was constructed for the sole purpose of rotary digging or cultivation.

In the following year a modified patent changed the power unit from a reciprocating engine to a rotary one and further alterations were made in July 1854, January 1855 and again in September 1856.

A complete unit was constructed and underwent trials in 1855. Romaine and John Henry Johnson worked together on the project, the final machine not surprisingly bearing a strong resemblance to Johnsons proposed machine of 1853 and in fact the patent was taken out in Johnsons name.

J. J. Mechi took a keen interest in Romaines machine and generously provided all the facilities for trials at his farm—Tiptree Hall in Essex, where unfortunately events proved the engine wanting in its capabilities.

In a lecture to the London Farmers Club in 1855 Mechi made the following remarks regarding his protegé: "As to his (Mr. Mechis') acquaintence with steam power, he believed that Mr. Romaines engine would have succeeded, and he had spent some money in making one; but upon trying it he found that there was too much philosophy to begin with. The revolving wheels performed 240 revolutions in a minute, and the earth and stones were made to fly in all directions; but in the process an amount of power was consumed which was

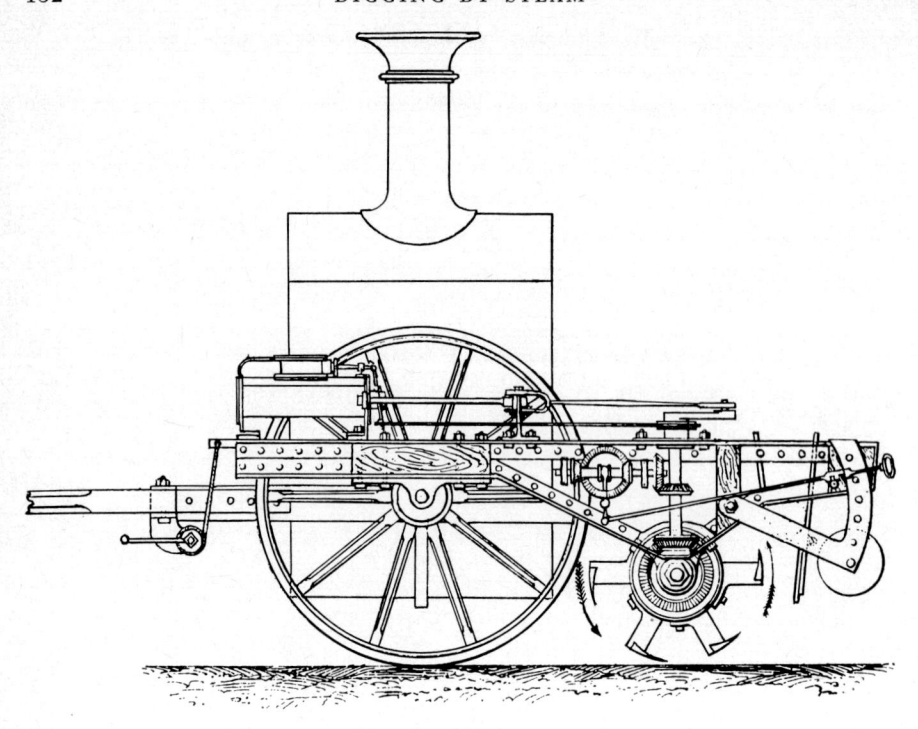

Fig. 103. Robert Romaine's cultivator of 1855. This horse drawn machine suffered from severe gearing and priming problems

altogether disproportionate to the quantity of land moved. There was not time for the prongs to get into the ground; and he really thought that with fifteen revolutions a minute he should have done better than with 240 (hear, hear). How far the engine would have the power of propelling itself he was unable to prove, because he adopted an improved system of boiler; and in order to get over the hills, the tubes were made vertical instead of horizontal; but with the high-pressure the first attempt sent the water and steam in a shower out of the boiler over the horses (a laugh). The boiler that would do a great deal of work in a horizontal form would, when the tubes were placed vertically, throw out the water with the steam, and soon become emptied. His experiment failed, therefore because there was too much philosophy, and too little steam (another laugh); and he did not continue it, inasmuch as it would have required both a large expenditure of time and an enormous outlay of capital; and it was certainly not his wish to spend another £500 upon it. He had heard, however, that Mr. Romaine had since made some experiments in Canada that were successful; but with most new inventions that they might anticipate failure at the outset, and they must be content if they advanced towards perfection by slow degrees (hear, hear).

With both vertical and horizontal boilers proving to be of little use from severe priming problems, it was back to the drawing board for Romaine.

Mechis' remarks—made in 1855—regarding Romaines return to Canada are something of a mystery as it was in that year that a model of an improved version of his design was shown at the Paris Exhibition. He must have returned to Britain for by 1857 a full size engine had been constructed. Made at the Beverley Iron Works under the supervision of Alfred Crosskill, this second digger possessed a twin cylinder 12 H.P. engine with the boiler still mounted on a two wheeled frame, but now with two wheels added for steering, and a fifth wheel at the rear, which ran on uncultivated land, regulating the depth of cut, equivalent to the land wheel of a plough.

The engine was of unpleasing proportions, looking somewhat like a foreshortened traction engine, and with the dispensing of the horses, was now capable of self movement at about 1 m.p.h. through a complex train of gears (Fig. 104). The steersman stood facing the rear on his own little platform at the front, his head dangerously near the flywheel the rim of which projected over the front of the engine.

The cultivator itself consisted of a drum constructed from boiler plates of 2 ft. 6 in. diameter. To the drum were attached the cultivating tines, and it could be lowered into work by means of a toothed rack and screw. Drive was by means of a shaft running from bevel gearing on the engine crankshaft, turning the cultivator at 40 to 50 r.p.m.

Although the cultivator width was greater than the track of the wheels, thus at least partially overcoming the problem of compaction, if only to the depth of the tines, there must nevertheless have been a panning of the subsoil.

Originally, the horse drawn version cost £600, but with the self moving engine it rose to a costly £800, due mainly to the complexity of the cultivating mechanism, although capable of being run quite economically, cultivating 4 to 7 acres per diem at a cost of 5/- to 10/- (25 to 50p) per acre.

W. H. Nash of Cubitt Town, Isle of Dogs made an engine to Romaines Patent, attempting to simplify its construction to enable it to be offered at a more reasonable price (Fig. 105). It was described by John Algernon Clarke in his prize essay on the "Application of Steam Power to Cultivation of the Soil".

The machine was 18 ft. long, 10 ft. wide and weighed 12 tons. The cultivator was 8 ft. wide, 53 inches diameter overall the tines which were 10 in. long.

Generally, the appearance was similar to Romaines engine of 1857. As with his engine the cultivator could be adjusted up and down in its frame, but a certain freedom was allowed to adjust automatically for irregularities in the soil surface the lower limit being set by the wheel that regulated depth of cut. In this way also, solid obstructions were overcome by riding over. Raising and lowering of the cultivator was by means of two cylinders and pistons to the rods of which were attached chains running to the hinged frame on which the cultivator rotated. Steam or water admitted to the cylinders adjusted the working height of the cultivator.

In seeing the machine working at Royston, Cambridgeshire, Clarke says: "on a piece of mown stubble—a light, chalky, turnip soil, of a quality that

STEAM CULTIVATION: CROSSKILL'S ROMAINE CULTIVATOR.

Fig. 104. Romaine's later rotary cultivating engine of 1857. Constructed by Alfred Crosskill, this public demonstration shows the engine apparently working well
(*Illustrated London News, 1857*)

Fig. 105. A further attempt to make an engine to Romaine's specification was made by W. H. Nash in 1859
(*British Museum*)

turns sticky with wet, but dry and friable at that time. The main wheels of 6½ ft. diameter, made of tee iron spokes, and plate-iron felloes 21 in. broad, sunk very considerably into this land, plainly showing the necessity for having flat platforms, like those of the endless railway (Boydell) over which the heavily weighted wheels might roll, but the difficulty of turning short round upon these shoes appears to be hardly overcome at present. The shares or cutters were arranged three in a row or ring, successive cuts being made on one revolution of the cylinder, and the advance for each cut being about 6 in. The digger made 30 to 40 revolutions per minute, the engine running at from 120 to 160 strokes per minute, varying according to slight elevations and depressions in the ground. The earth was well subdivided by the 33 cutters (that is in 11 rows) and turned up in small spits, beautifully broken, much like spade work, the depth not exceeding 6 in.; or rather 6 in. at one end of the digger and 4 in. at the other, owing to a want of rigidity in the lever framing. That it is proposed to remedy by partly bearing up the end which is not supported by the small travelling wheel by means of the hydraulic jack, and this wheel also is to be placed exactly opposite the end of the digging cylinder, so as to adjust the depth of the culture more accurately. In deeper tillage the action of the machine is far more effective and economical, and it is able to dig 12 and even 16 in. deep. A length of 23 chains was cultivated in 23 minutes, including 1½ minutes stopping, turning and setting in again; and as the breadth of the work is 7½ ft., the extent dug in a day of 10 hours is nearly 7 acres.

The engine, working at 70 lb. pressure, may be called 20 horse power, though it may be worked up to 80 or 90 lb. if required, and hence the expenses may be estimated. These worked out at between 6s. (30p) and 6s. 6d. (32½p) per acre". Rather over enthusiastically he mars a remarkable account by going on to say that "This is extraordinary economy, when we consider the effective character of the tillage, and that work 10 or 12 in. deep would cost but little more; that, is merely extra fuel for the greater power engaged". Clarke making the mistake of thinking that twice as much coal equals twice as much power!

The cost of Nash's engine was £700—somewhat cheaper than Romaine's.

Romaine did make several more rotary diggers referring in a letter to *The Engineer* in 1862 to "my last six digging machines".

In spite of Mechi's encouragement and the practical assistance of others, Romaines attempts at rotary digging came eventually to nought.

James Howard reported in 1862 that Romaine had abandoned his rotary engine in favour of "rope traction" and apparently took out "one or two" patents.

Mechi had shown a model of Romaines horse drawn cultivator to the Society of Arts in 1853. Hopefully he said "If it does not supersede the plough, it will limit its operations. When once the steam cultivator is shown to answer no doubt many others will appear and I venture to predict that, within seven years, steam will become the grand motive cultivating power".

Prophetic words not to come true in Romaines case or any other digging machine worker, but to come true at least partially, within the prophesied time by the efforts of John Fowler, William Smith and others, for by the time seven years had elapsed, there were indeed many inventors, none of whom had completely solved the technical and financial problems, and all working on different principles to Romaines machines.

C. W. Hoskyns

As noted earlier, Chandos Wren Hoskyns was very active in promoting steam cultivating engines and was probably the only one to advocate both the rotary and spade principles of digging.

In his book *Talpa: or the Chronicles of a Clay Farm* the farm journalist scorned the principle of horse ploughing, maintaining that the horse was merely another form of animal power merely used to tow a plough in the only way possible—horizontally and in Hoskyns opinion, inefficiently. In 1849 in his *Inquiry into the History of Agriculture*, he commented that "The application of steam power to tillage—will probably be vastly more successful if made to imitate the action of the spade".

"The idea of the plough in this subject is a misleader. What we want is, not to plough the land, but to cultivate it; and if the plough and all its subsidiary implements are a mere substitute for the spade, and on stiff soils in a moist climate a very expensive, cumbrous and imperfect one; the object of the inventive machinist will be better directed as well as simplified, by discarding it altogether from his thoughts and concentrating his attention to the action of the spade."

To use the new found power of steam merely to replace the horse was entirely against Hoskyns ideas. He hated the "employing of a Steam-engine to turn a Drum, to wind a Rope, to drag a Plough, to turn up a Furrow".

"Get into steam-power and you have no more to do with the plough, than a Horse has to do with a spade. It is no *essential whatever* of cultivation that it should be done by *the traction of the implement*. Spade work is perpendicular.

Horse work is horizontal. Machine work is circular. . . . The steam engine has no taste whatever for straight draught. He is a revolutionist, in the most exact sense of the word. He *works* by revolution: and by revolution only will he cut up the soil into a seed-bed, of the pattern required, be it coarse or fine."

Hoskyns explosive book tore to shreds traditional ideas, condemning "all the train of piebald monstrosities and biform incongruities that mark those periods of false gestation and miscarriage in the annals of steam ploughing inventions".

Putting his words into practice, in 1853 Hoskyns' produced a patent for an engine which was somewhat similar in appearance to Robert Romaines machine, patented in the same year (Fig. 106).

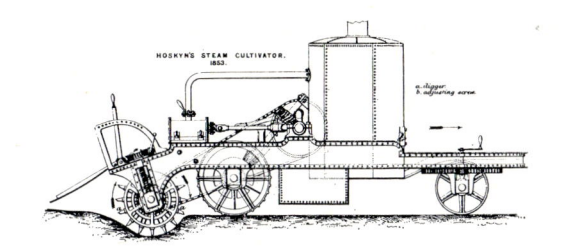

Fig. 106. Steam cultivator of 1853 by C. W. Hoskyns
(*Transactions of the Society of Engineers 1868*)

In contrast to Hoskyns' earlier propounding of spade digging he had designed a rotary digging engine which was "applied in such a manner as to accomplish at one process the due preparation of the soil for a seed bed. This result is obtained by a machine which instead of lifting in mass the stratum of earth under cultivation, like the plough or spade is so applied to the soil as to reduce it by abrasion to the required tilth in fineness. This abrasion is performed by a series of discs or wheels, fixed on a rotating axis actuated by steam power, the periphery of which discs are furnished with radiating points or cutters. The rotary motion of the disc is communicated from the steam engine, from which also the progressive motion of the machine is derived. The two motions are independent of each other, and so arranged that a rapid motion may be given to the cutters, while the progressive motion is slow or suspended altogether, as at commencing. The gearing is such that the respective speeds can be varied at pleasure to suit the nature of the soil. The cutters, by their rotary action, first enter the soil, making a semicircular trench, which, during the progress of the machine, is constantly preserved at the required depth, and the soil, abraded and cut down as the machine advances, is thrown off tangentially behind in a comminuted, inverted and aerated condition".

The engine was a vertical boilered twin cylinder machine. A heavy iron framework was equipped with two large rollers at the rear and two smaller ones at the front. The large rollers were "straked" with raised strips at right angles to the roller giving an improved grip on the soil. Drive to the rollers was by means of worm and wormwheel from the crankshaft via an angled shaft from the engine. The roller end of the shaft was driven by bevel gears and positioned in the middle of the framework. Both ends of the crankshaft were provided with large sprockets which revolved the cultivating cylinder through chains. This cylinder or drum was equipped with spikes around its circumference and along its length, and in biting into the ground, made a trough which was progressively "rasped" away as the whole machine moved forward, the soil thus loosened being deposited at the rear of the drum, having been finely broken and tumbled about achieving the desired "comminution" and "aeration".

Hoskyns' machine, popularly named his rasping or scratching engine, was probably never built as the author has been unable to trace any evidence of trials. It is probably just as well for it would have been better had Hoskyns remained a journalist. His engine was mechanically of bad design, it depended upon the soil being of a sort and in such a condition that it would not clog on the one hand or break the teeth of the rasp on the other, thus severely limiting its application.

Similarly to Ushers machine, the steering was very flimsy and provided merely as an afterthought, or so it seems. A nearly complete gear ring the width of the engine meshed with two very small pinions with equally small, blister raising handles, about 9 in. long, and intended to steer 10 tons of engine!

Hoskyns' steam rasp was unsuccessful, but others followed, emulating his rotary principle, attempting to improve upon it. His volatile writings possibly did more to promote the idea of steam digging than any other single individual and certainly did more for his cause than his rather inadequate engine.

The patent specification gave a rate of advance of 200 yds. per hour leaving a rasped swathe 9 ft. wide. This gave an approximate area of $1\frac{1}{4}$ acres per day. Quite small when one considers the first cost of the engine and the expense of running it, and the great amount of energy required to cultivate the small area made the steam rasp unacceptable not only on financial grounds, but also from the mechanical point of view, both of which ensured the speedy demise of Hoskyns' schemes.

CHAPTER TEN

The Archimedians

ONE SYSTEM OF ROTARY steam cultivation was to utilise archimedian screws. These were intended usually to fulfil dual functions, one to disturb the soil and the others to propel, or assist in propelling the complete engine along, at the same time.

Apart from the engineering and technical difficulties of producing an engine of this type, the potential hazards are clear when one considers—for example—the varying textures and consistency of the soils—some with stones and others containing sticky clay.

Nothing daunted, several inventors proposed engines which worked in this principle, among them being Holcroft, Monckton and Clark, Henry Parker and John Brennand.

Holcrofts agricultural steam engine of 1856 (Fig. 107) relied for propulsion entirely on a huge archimedian screw which bit into the soil, and when revolved, dragged the engine along. This engine was not intended for cultivating, but for "driving machinery, for manufacturing drainage pipes, threshing grain, or other purposes". It is difficult to speculate why Holcroft did not include in his patent the additional function of ploughing, as the impractical looking engine would have automatically left a dug, semicircular groove in its wake wherever it went—and incidentally, would have been unable to travel that

Fig. 107 Henry Holcroft's ingenious but impracticable design for a screw propelled engine
(*The Engineer*, 1857)
(*M.E.R.L.*)

particular path again until the soil had compacted sufficiently to obtain a "bite"! It is even more incomprehensible that Holcroft did not attempt to drive his engine along by means of chain or shaft drive. Such are inventors whims!

Assuming for one moment that there was a grain of practicability about this engine, it was logical that the archimedian screw should be modified by attaching plough shares, coulters or spikes around the rim and the engine directed backwards and forwards across a piece of land to cultivate it. This is indeed how some inventors intended their engines to work. None did though, on this principle, but some very attractive drawings are left to us to show how the engines would have appeared.

In the same year as Holcrofts proposals, Messrs. Monckton and Clark patented an engine, also vertical boilered but horizontally twin cylindered (Fig. 108). The whole engine was supported on two rollers, each driven by bevel gears from a horizontal shaft which possessed a crank at each end. The cranks provided drive to the strange looking cultivating screw which was positioned at right angles to the rollers, running along the side of the engine. As the engine proceeded along, the cultivating screw carved a semi-circular groove in the soil. The screw was formed of many tines arranged in spiral fashion around its central shaft, shorter tines at each end increasing in length

Fig. 108. As unlikely a design as Holcroft's, the strange engine of Monckton and Clark's cultivated grooves in the soil while travelling sideways. Delightfully "cranky", at least this effort of 1856 attempted to solve a technical problem, while many others were willing to criticize, without themselves becoming involved in the expense of such ventures
(*The Engineer, 1857*)
(*M.E.R.L.*)

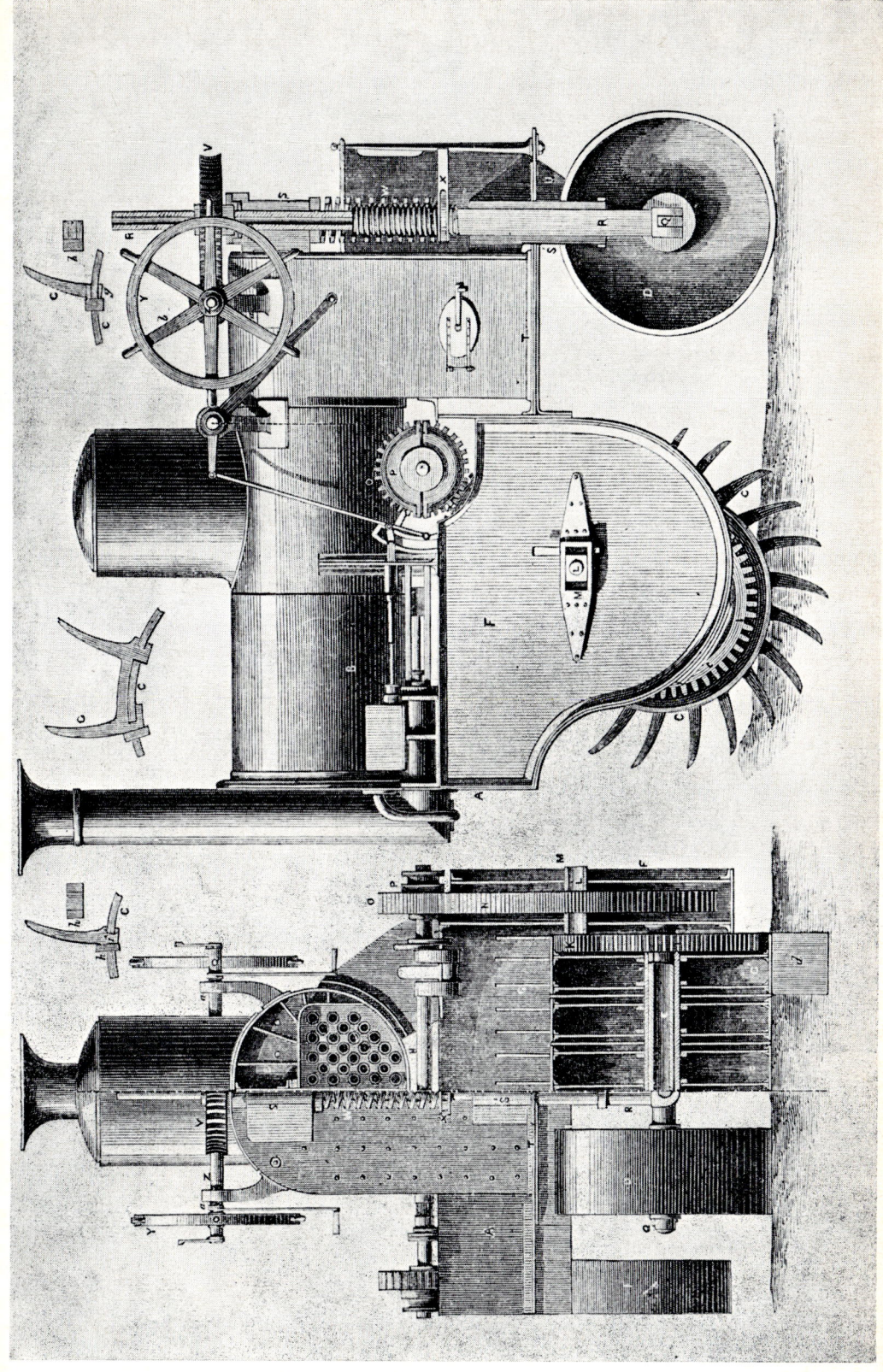

Fig. 109. Henry Parker's engine of 1858. The cultivator also propelled the engine along
(*The Engineer*, 1859)
(*M.E.R.L.*)

Fig. 110. John Brennand's engine of 1858 was among the last attempts to dig by means of a self propelling cultivator. Some years were to elapse before more sophisticated engines working on more practical principles appeared
(*The Engineer*, 1859)
(*M.E.R.L.*)

to a maximum in the middle, enabling the cultivating cut to be made in either direction. A "comb" was positioned along the screws length to clear the clogging soil and stones away from the tines. How the engine was steered is not apparent from the engravings of it.

In October 1858, Henry Parker patented his engine for rotary cultivation of the land which consisted of a more or less standard traction engine boiler with cylinders mounted underneath (Fig. 109). Two narrow tracked front wheels provided steerage, while the wide single roller at the rear was equipped with a great number of small curved "paddles" which on biting into the soil, not only pulverised it, but provided a degree of propulsion. As the speed of revolution of the roller was at the rate of progress of the engine, it would seem that the pulverisation was brought about almost entirely by the weight of the engine forcing the "paddles" into the ground. The main bulk of the engine was positioned directly over the roller axle to help penetrate as much as possible. This was a rather ungainly and impracticable enterprise which did, however show some improvement over the pure archimedian principle of propulsion and cultivation.

John Brennand of Manchester proposed in his patent of 1858 a vertical boilered engine mounted in an iron frame supported by four small waggon type wheels, the front pair having horse shafts attached (Fig. 110). The horse would have served a steering function and may also have assisted in hauling the engine, supplementing the inadequate forward movement provided by the screws.

No record survives of any trials of this engine, if indeed it was ever made. Brennands machine was among the last to be designed on this particular rotary principle, it being found wanting in many respects, not the least being the inconsistency of the soil.

After 1860 other means of cultivation to compete with the steam plough were sought. Improvement not only in basic design work but in the power unit was essential before any hope of success in cultivating the soil could be entertained. Too much concentration in the theoretical aspects of what *should* work, as opposed to practical trials of what *would* work resulted in failure of all the rotary machines of the 1860s.

Underpowered, flimsy cylinders, motions and sometimes boilers, crankshafts often without flywheels, occasionally top heavy or over complicated resulted in much time and money being spent, all of which came to nought.

CHAPTER ELEVEN

Deeper Diggings

THE FOLLOWING PAGES are a condensed version of some facts and figures, with some additional details included, enabling reference to be made regarding information on digging engines which may not be speedily found otherwise.

Broadside digging engines.

The fabled broadsiders were but few in number. Listed below are all details known of the twenty-one or so engines of this type constructed. All possessed single cylinders.

Maker	Engine No.	Order Date	H.P.	Weight Tons	Sold To	Date	Remarks
T. C. Darby and Eddingtons			8	9	Darby	1877	First walker
T. C. Darby and Eddingtons			8	9	Darby	1877	Rebuilt walker
T. C. Darby and Eddingtons			8?	9?	Darby	1878	Scrapped
T. C. Darby and J. & H. McLaren			10		Darby	1879	Last walker, first double boiler
T. C. Darby					Darby	1879–80	Development only, first wheeled m/c
T. C. Darby and Eddington?			10?		Darby	1880	First full size wheeled m/c
J. &. H. McLaren	82		8	15t 7c	Darby	1880	Used at Pleshey
J. & H. McLaren	133		8	15t 7c	Wimshurst, Hollick	1881	Fate unknown
J. & H. McLaren	134		8	15t 7c	Wimshurst, Hollick	1881	Fate unknown
J. & H. McLaren	135		8	15t 7c	Wimshurst, Hollick	1883	Fate unknown
F. Savage	399	1887	8	15t	J. W. Moss	1888	"The Enterprise"
F. Savage	429	1888	8	15t	Hall	1888	Fate unknown
F. Savage	436	1888	8	15t	Harrison or J. Walker	c1890	Possibly the "Juggernaut"
F. Savage	469	1889	6	9t 5c		c1891	One surviving photograph

Maker	Engine No.	Order Date	H.P.	Weight Tons	Sold To	Date	Remarks
F. Savage		c1895	6	9t 5c	Not sold		Last broadsider. Broken up.
The Agricultural & General Engineering Co.			8			1881–82	Little detail known

There is the possibility that Darby made up to five further broadsiders.

All McLarens engines possessed cylinders 9 × 12 in., Savages cylinders were $9\frac{1}{8}$ × 12 on the 8 H.P. and 8 × 10 on the 6 H.P. engines.

Digging engines made

The following list is intended to show the approximate number of digging engines known to have been constructed. Many digging exponents ideas remained on paper, never reaching constructional actuality for the reasons mentioned in earlier chapters. No doubt a few are missing from the list but it is as complete as possible in the light of information available. Totalling some one hundred and fifty-five engines and models, they are the sum result of incalculable amounts of time and money.

Reciprocating digging engines

Maker	Number made	Year	Remarks
T. C. Darby	1 or 2	1877	Models
	10 to 12	1878–90	Broadsiders. (See list)
	1	1976	2 in. scale model by C. R. Tyler
Darby Syndicate	30 plus	c1890	Quick Speed Digger
Darby-Steevenson	1	1890	Made by Savages. Works No. 507
	5	1890–93	Colchester and Nottingham diggers by Davey Paxman
Cooper Steam Digging Co., Ltd.	1	1891	Converted traction engine
	1	c1893	"Pioneer"
	1	1894	Twin cylinder digger
	5	1894 on	Made by Fowler
	36 plus	1900–10	Number uncertain
	approx. 24	1903	No. 5 engine, mostly exported to Egypt
Frank Proctor	approx. 12	1886 on	Made by Burrells. Mostly exported to Europe
J. H. Knight	1	1874	$\frac{1}{2}$ scale model
	2	1874	Made by Hetherington and Parker, Alton
	2	1876–77	Made by Howards
J. D. Garratt	1	1879	Model
	1	1883	Made in Germany

Rotary digging engines

Maker	Number made	Year	Remarks
T. Atkins	1	1843	Model to Atzlars patent
	1	1853	Full size engine

Maker	Number made	Year	Remarks
P. P. C. Barrat (France)	10–20	1847	Possibly the first successful digging engine made
Calloway and Purkis	2	1851	Models
James Usher (Scotland)	1	1850	Model. Exhibited today at Science Museum, London
	3	1851 on	First one by Slight. Possibly other two as well
Thomas Rickett	2	1858–59	Second one 15 H.P.
John Bethell	1	1852	Modified portable
	1	1857	Purpose built digger
Robert Romaine	1	1853	Model horse drawn machine
	1	1855	Full size horse drawn steam digger
	1	1855	Model shown at Paris
	1	1857	Made by Crosskill. Self moving engine
	1 plus	1859	Made by W. H. Nash
	6 plus	c1859	Possibly Crosskill type but could be Nash type

List of Digging Engine Patents (Selection)

Patent No.	Year	Patentee	Subject
3309	1810	Major Pratt	Spiked digging machine
	1843	Atkins/Atzlar	Digging engine
11297	1846	Bonsor and Pettit	Rotary cultivator
11907	1847	Sir John Scott Lillie	Rotary cultivator
12710	1849	James Usher	Rotary cultivator
1899	1851	Chandos Wren Hoskyns	Land Rasp
949	1852	John Bethell	Rotary cultivator
	1853	Robert Romaine	Rotary digger
1151	1853	John Henry Johnson	Rotary Tiller
260	1854	Atkins/Atzlar	
35	1855	John Henry Johnson	Digging machine
2065	1856	E. H. C. Monckton	Machine for tilling the land
2552	1856	Henry Holcroft	Agricultural steam engine
991	1857	Alfred Vincent Newton	Machine for cultivating land
	1857	Thomas Rickett	Implements for cultivating
2194	1857	Thomas Keddy	Machinery for cultivating land
2691	1857	John Bethell	Boydell wheeled engine
1022	1857	John Blythe Robinson	Apparatus for effecting agricultural operations
1849	1858	Thomas Rickett	Locomotive engines
	1858	John Brennand	Improvements in agricultural engineering
2051	1858	Parker	Apparatus for cultivation of the land

Patent No.	Year	Patentee	Subject
612	1862	Fowler, Greig and Noddings	Cultivating by rotovator
3797	1873	J. H. Knight	Steam digging machine
4942	1876	J. H. Knight	Steam digging machine
51	1877	T. C. Darby	Cultivating apparatus
2555	1877	T. C. Darby	Cultivating apparatus
1773	1879	T. C. Darby	Digging machinery
4926	1881	T. C. Darby	Digging machinery
	1884	Proctor	Apparatus for cultivation of the land
5956	1884	M. R. Pryor	Digging mechanism
14608	1886	D. Nagy (Budapest)	Improved steam plough
9059	1886	Proctor, Burrell	Combined traction engine and digger
8385	1889	J. Johns	Apparatus for tilling the land
11963	1893	T. Cooper	Improvements on digger
23606	1897	B. P. Nubar (Cairo)	Machine for ploughing and cultivating the land
22570	1898	W. J. Burgess	Steam cultivators

Acknowledgements

When researching historical matters connected with steam engines, it is invariably the case that assistance sought is willingly given. No doubt other writers on different subjects find similar help, but one wonders if the common interest of steam perhaps produces a spirit of co-operation which is unique. Enthusiasts combine to pool common interests and to a certain extent this book is the product of specialised pieces of knowledge gleaned over several years from individuals who were most helpful in all cases.

Acknowledgement for this help is made here to all those of whom I am aware. Should any not be mentioned that ought to be, my thanks and apologies.

John Booker (Essex Record Office)
Ronald H. Clark
John Creasey (Museum of English Rural Life, Reading)
G. F. A. Gilbert (Road Locomotive Society)
Anthony S. Heal
Arthur Johnson
Rev. R. C. Stebbing
R. Trett (King's Lynn Museum)
B. M. Taylor

Record Offices, Museums, Libraries and Companies
British Museum
Coopers Roller Bearings, Ltd.

Essex Record Office
G.E.C. Diesels Ltd., Paxman Process Plant Division
The Lynn Museum
Museum of English Rural Life, Reading University
Road Locomotive Society
Savages of King's Lynn
Science Museum (London)
Science Reference Library

Magazines and Newspapers
Eastern Daily Press
Engineering
Illustrated London News
Illustrations
Model Engineer
The Engineer

Manuscripts, Catalogues, etc
International Exhibition Catalogue 1862
Anthony Heal lecture to the Road Locomotive Society
J. A. Clarkes letter to the Royal Agricultural Society 1859
The Patent Office—various patents
Manufacturers catalogues:—
Coopers Roller Bearings, Ltd.
Darbys Patent Pedestrian Broadside Digger Co.
J. H. Knight
Ransome, Sims and Jeffries
F. Savage

Bibliography

During the course of writing this volume, references have been made to various publications. I believe that most sources of steam digging information are included, but there must, of course, be some details yet to be discovered. I would therefore be grateful to learn of any details of steam diggers or anecdotes connected with them which would assist my researches further.

As far as I am aware, permission for reproduction of information and illustrations has been sought and in every case granted. I tender my grateful thanks to the publications concerned and my apologies to any that may have been omitted.

Journals

Journal of the London Farmers Club 1854–64.
Journal of the Road Locomotive Society.
Journal of the Royal Agricultural Society of England.
Journal of the Society of Engineers 1868.

Books

An Inquiry into the History of Agriculture. C. W. Hoskyns, 1849.
Development of the English Traction Engine. R. H. Clark, 1960.
Essex and the Industrial Revolution. J. Booker, 1974.
God Speed the Plow. Clark C. Spence, 1960.
Great Farmers. J. A. Scott Watson, Mary E. Hobbs.
Ploughing by Steam. John Haining, Colin Tyler, 1970.
Richard Trevethick Memorial Volume. H. W. Dickinson, Arthur Titley, 1934.
Steam Engine Builders of Cambridge, Sussex and Essex. R. H. Clark, 1950.
Steam Engine Builders of Lincolnshire. R. H. Clark, 1955.
Steam Engine Builders of Norfolk. R. H. Clark, 1948.
Talpa, or the Chronicles of a Clay Farm. C. W. Hoskyns, 1853.
The Art of Husbandry. J. Mortimer, 1708.
The Same Sky Overall. David Smith, 1948.

Index

Agricultural and General Engineering Co.   45, 50, 56, 57, 92
Alton   117
Atkins, T.   133, 134
Atzler, J. A.   133, 134
Aveling and Porter   92, 144

Badshot Farm   117, 119
Bailey, Mrs.   88
Baguley Cars Ltd.   89
Barrat Bros.   135
Barrat, P. P. C.   41, 135
Bath Show   119
Bauer   137
Beadel, W. J.   61
Beards of Stowe   148
Belmec International   78
Bentall   89
Bessemer   23
Bethell, John   149 et seq
Beverley Iron Works   153
Boydell   155
Brennand, John   159, 163
Bristol and Islington Cattle Show   45
Brooman   137
Buckau   105
Burrell   64, 83, 92, 109, 110

Calloway and Purkis   106, 137
Calloway, George   137
Carlisle R.A.S.E. Show   51, 53, 54, 61
Chambers, Charles   116, 119
Chelmsford Station   58
Chester R.A.S.E. Show   145, 146
Clark, Ronald H.   57
Clarke, John Algernon   112, 113, 146, 151, 153
Clayton and Shuttleworth   83, 92
"Colchester" digger   87, 88, 89
Coleman and Moreton   36
Cooper Roller Bearings Co. Ltd.   105

Cooper Steam Digger Co. Ltd.   29, 74, 93 et seq 108
Cooper, Thomas C.   93 et seq
Crosskill, Alfred   153
Crystal Palace   139

Dalby, Prof. W. E.   97
Darby, John   34
Darby Land Digger Syndicate   80, 81, 83, 92
Darby-Maskell   89
Darby Patent Pedestrian Broadside Digging Co. Ltd.   45, 61, 62, 90, 92
Darby, Sidney   37, 92
Darby, Sidney (Wickford) Ltd.   83
Darby-Steevenson   44, 73, 87, 89
Darby testimonial fund   55
Darby, Thomas, Snr.   83
Darby, Thomas Churchman   29, 31, 34 et seq 67, 68, 73, 75, 80, 92, 93, 105, 121
Darby workshops   44
Davey, Paxman and Co.   87, 89, 92
Digger hire   61
Duke of Buccleuch   88
Dussuc   137

Eddington and Darby   40
Eddington and Steevenson   44, 87
Eddington, Sylvanus   36, 44
Eddington, William   44
Eddington, W. & S.   36, 37, 41, 44 et seq 64, 92
"Engineering"   117
"Engineering Reminiscences"   117
Essex Agricultural Society   110
Essex and Chelmsford Museum   47

Fairlie locomotive   35
Farmers foundry   93
Feering,   64, 66
Festiniog railway   35

171

Field, Justice 55
Fisken Bros. 124, 125
Fowler, John 30, 83, 92, 97, 127, 156

Garrett and Co. 93, 116
Garrett, John D. 105
Garrood, Charles 108
Grand Duke of Baden 109
Great Caulfield 88
Great Ryburgh 93
Gyrotiller 31, 90, 138

Halls of Dartford 131
Harrison Bros. 66
Hawkins, Sir Christopher 129, 131
Heathcoat, John 133, 135
Hetherington and Parker 117, 119, 121, 125
Highland and Agricultural Society 139
Hilliard, G. B. 61
Holcroft, Henry 24, 159, 160
Hoskyns, Chandos Wren 25, 156 et seq
Howard, James 134
Howards of Bedford 35, 119, 121, 127

International Exhibition 139

Jasper 130
Jeannevet 137
Johnson, Arthur 69, 72, 73, 79, 80
Johnson, John Henry 151
Johns, J. 106

Keddy, Thomas 137, 138
Keddy's Patent Traction Engine, Steam Plow and Land Cultivation Co. 137, 138
Kendal 130
Kelvedon 64
Kilburn Show 40, 45, 47
Kings Lynn 60
Kings Lynn Museum 79
Knight, John Henry 34, 110, 111, 114 et seq

Langley Park Farm 119
"la piocheuse" 135
Layer Marney 64
Lillie, Sir John Scott 24, 135
London Farmers Club 134, 142, 141
London Science Museum 130, 139
Lord Dedunstanville 130

Marquis de la Laguna 109

Mason 88
Maudsley, Henry 23
Mechi, J. J. 34, 142, 151, 153, 156
Monckton and Clark 159, 160
Mortimer, J. 20, 92
Moss, J. W. 64, 70, 86, 87
"Mud Show" 61
Murdock 22
McAlpine, W. 79
McEwan Pratt 89
McLaren, J. & H. 40, 45, 51 et seq 61, 63, 64, 73, 92

Nagy, D. 106
Nash, W. H. 153, 154, 155
Nasmyth 23
Newcomen 22
New Street Ironworks 44
"Nottingham" digger 87, 88, 89
Nubar, B. P. 107

Parker, Henry 111, 159, 163
Patent Office 41
Pedestrian digger 45
Phillips, G. C. 31, 61, 83
Philosophical and Natural History Society 47
"Pioneer" 93
Pleshey Lodge Farm 34, 36, 39, 47, 56, 60
Pratt, Major 113, 132, 133
Preston R.A.S.E. Show 54
Proctor, Frank 108, 109 et seq 119
Pryor, M. R. 111
Purkis, Robert 137

"Quick Speed" digger 74, 82 et seq 89, 97, 100, 108

Ransomes, Sims and Jeffries 30, 83, 84, 92
Rastrick 130
Rickett, Thomas 143 et seq
Roberts, M. J. 133
Romaine, Robert 151 et seq 157
Royal Agricultural Society of England 28
Ruston-Proctor 83, 92

St. Nicholas Iron Works 67
Salisbury R.A.S.E. Show 115
Savage Darby 6 H.P. digger model 75 et seq

Savage, F. Ltd.   35, 41, 56, 63 et seq 89
Savage, Frederick   66, 68, 69, 70, 74, 80, 113
Savory   22
Scholes   55
Sible Hedingham   56
Shrewsbury R.A.S.E. Show   133, 134
Sinclair, Sir John   131
Slight of Edinburgh   139
Smithfield   50
Smith, William   41, 156
Society of Arts   134, 150, 156
Steam Spade Tormentor   128
Stebbing, Rev'd. R.C.   70
Steevenson   44
Stilemans Works   83, 85, 87, 92

Terrington St. John   64, 66
Tiptree Hall   151
Thaxted   56
The Castle Foundry   143, 148
"The Caulfield Enterprise"   88
"The Engineer"   148
"The Enterprise"   63, 64, 69, 70, 72
"The Juggernaut"   66
The Model Engineer   49, 75
Trevethick, Richard   22, 128 et seq

Usher, James   139 et seq 158

Waghausel   109
Walker, James   66, 69
Walking diggers   37
Wallis and Steevens   83, 92
Warwick R.A.S.E. Show   147
Wey Iron Works   121
Whatley   147
Wickford   85, 92
Wigborough   64
Williams, J. A.   141, 142
Wimshurst, Hollick and Co.   51
Wright, Walter   55

York R.A.S.E. Show   97
Yorkshire Steam Wagon   35

Zichy, Count Andor   83, 88